D1006088

WATER *to the* ANGELS

ALSO BY LES STANDIFORD

Desperate Sons

Bringing Adam Home

The Man Who Invented Christmas

Washington Burning

Meet You in Hell

Last Train to Paradise

THE JOHN DEAL SERIES

Done Deal

Raw Deal

Deal to Die For

Book Deal

Deal with the Dead

Presidential Deal

Bone Key

Havana Run

OTHER NOVELS

Spill

Black Mountain

WATER *to the* ANGELS

William Mulholland, His Monumental Aqueduct, and the Rise of Los Angeles

LES STANDIFORD

ecco

An Imprint of HarperCollins*Publishers*

Credits for insert photographs: pages 1, 3 (*bottom*), 5 (*top*), 6, 8 (*top*), 9 (*bottom*), 12, 13 (*bottom*): courtesy of the Los Angeles Department of Water and Power; pages 2, 3 (*top*), 4, 5 (*bottom*), 7, 8 (*bottom*), 9 (*top*): courtesy of the County of Inyo, Eastern California Museum; page 10 (*top*): Los Angeles Public Library—Los Angeles Herald Collection; pages 10 (*bottom*) and 11 (*top*): H. T. Stearns, USGS; page 11 (*bottom*): Ventura County Library; page 13 (*top*): George R. Watson, Watson Family Photo Archive; page 14: WikiCommons; page 15: courtesy of Clinton Steeds; page 16: Courtesy of the Library of Congress, Jet Lowe

FIRST EDITION

Designed by Suet Yee Chong
Map of the Los Angeles Aqueduct from the Report from the City of Los Angeles, 1971

Library of Congress Cataloging-in-Publication Data has been applied for.

ISBN 978-0-06-225142-8

15 16 17 18 19 OV/RRD 10 9 8 7 6 5 4 3 2 1

To James "Bimmy" Wantz, who opened my ears to the siren song of California so many years ago

And to Mitchell Kaplan, whose dedication to the enterprise of bookdom is legend

Like a man gone out of Egypt . . .
Before me the desert,
Perhaps the Promised Land, too.

—YEHUDA AMICHAI

The mysterious is the source
Of all true art and all science.

—ALBERT EINSTEIN

CONTENTS

ACKNOWLEDGMENTS

My sincere thanks go to a number of individuals and institutions without whose help this undertaking would have not been possible:

Holli M. Lovich, Special Collections and Archives Coordinator for the Oviatt Library, California State University at Northridge, and Project Archivist for the Catherine Mulholland Collection

Dr. Paul Soifer, Consulting Historian, Los Angeles Department of Water and Power

Fred Barker, Waterworks Engineer and unofficial department historian, Los Angeles Department of Water and Power

Heather Todd, Archive Gatekeeper for the Eastern California Museum in Independence, California

Angela Tatum, Office of the Los Angeles Department of Water and Power Archive

The staff of the Huntington Library, San Marino, California

Adis Beesting, Education Librarian, Green Library, Florida International University

Marissa Ball, Emerging Technologies Librarian, Florida International University Libraries

Beatriz F. Fernandez, Reference Librarian, Green Library, Florida International University

Christine Mulholland, great-granddaughter of William Mulholland and niece of Catherine Mulholland. Her grandfather was William Perry Mulholland, William's first-born son.

Harold "Hal" Eaton, great-grandson of Fred Eaton.

Douglas Wartzok, Provost and Executive Vice President, and the Sabbatical Leave Committee, Florida International University

I would also be remiss if I failed to thank those who have encouraged me in this work from the very outset: my steadfast agent, Kim Witherspoon, and my undaunted editors, William Strachan and Daniel Halpern.

Thanks are also due to Bill Beesting for his close reader's eye, to Meg Grant and Greg Lecklitner for their untiring and ever-gracious Los Angeles hospitality, to my colleague James W. Hall for his unflagging encouragement and perspicacity, and of course to my ever-supportive wife, Kimberly, who has enabled the habit of an ink-stained wretch for more than thirty years now.

AUTHOR'S NOTE

Often a writer is queried as to the source of an idea. Ordinarily it is an impossible question to answer. But in this case lies an exception. Whatever one's opinion of upstart Los Angeles—and there are many—there is no arguing that it stands as a major population center of the world, along with the likes of Tokyo, Mexico City, Rome, New York, Cairo, Delhi, or Beijing. But for a moment try to comprehend that one individual could be given credit for the existence of any of those cities named. If such a concept is intriguing, then one might wish to read on.

When this writer first ventured to Southern California, it was from a dimly lit existence in southeastern Ohio. There were hills there on the fringes of Appalachia, but they overlooked for the most part only other hills. And at night, there was little to be seen from any ridge but an endless expanse of ruffled indigo.

Imagine then, a young man fresh from the sticks being driven along the spine of the Hollywood Hills as the sun sinks into the Pacific to the south and dusk sweeps across the San Fernando Valley to the north. At a turnout off a narrow road as winding as any to a coal miner's shack, standing beside a gas-guzzling sedan, engine off and pinging beneath the hood in the dark, the young man stares out across an endless valley paved with a carpet of lights that stretch to infinity. A real-life fairyland from this vantage point, and don't bother trying to tell him any different.

"What's the name of this road we're on?" the young man wants to know.

"Mulholland Drive," his companion says. "What do you care?"

"Just so I know how to get back here," the young man says, gaping at the view below. Whoever Mulholland was, he must have been important. It's one of the most amazing roads in the world.

IN SOME RESPECTS, of course, it is a fool's errand, trying to excavate grandeur out of the past. Every day, thousands of people drive along or across Mulholland Drive, the familiar portion of which traces the crest of the Hollywood Hills in Los Angeles, all the way from the 405 Freeway to Highway 101. It is a way to get somewhere or a place to enjoy and the name is just something to plug into Mapquest or a smartphone. Who should think about William Mulholland? Why would anyone care?

Catherine Mulholland, granddaughter of the man, grappled with the matter in significant ways. In an unpublished bit of autobiography she remembers often being asked, "Are you related to the highway?"

"And when I reply that it was named for my grandfather in 1925, the response ordinarily produces neither recognition nor much curiosity. In high school I sometimes longed for such indifference, because in the late 1930s, Mulholland Drive had become the foremost trysting, parking and necking spot in Los Angeles."

It turns out that William Mulholland was in fact the highest-paid public official in California a century or so ago, but by today's standards that doesn't mean much—that grand annual salary wouldn't cover the mortgage payment on most of the homes along the road bearing his name today. In 1974, Robert Towne wrote the script for *Chinatown*, an acclaimed film that derives substance from some of the things Mulholland was involved with, but he is called Hollis Mulwray in the film, his fictional character is minor and killed early on, and the filmmakers had to fudge most of the actual facts—including the time frame—for fear audiences would get lost in the forests of history.

In the latter half of the twentieth century, a number of factually based works were written regarding the fierce water politics in the West, including Marc Reisner's notable *Cadillac Desert: The American West and Its Disappearing Water* (1993). Mulholland and the Los Angeles Aqueduct get their fair share of mention there and in other books, including William Kahrl's exhaustive *Water and Power: The Conflict over Los Angeles' Water Supply in the Owens Valley* (1983) and Abraham Hoffman's *Vision or Villainy: Origins of the Owens Valley–Los Angeles Water Controversy* (1981), which make the aqueduct project their primary focus.

However, as some—including Catherine Mulholland—have lamented, many of the water-issue books are dominated by political concerns and a seemingly unavoidable tendency to side with one group or another of the "good guys" du jour. "I estimate that of the voluminous literature which exists concerning the history of Owens Valley and water in Southern California," Ms. Mulholland once stated, "that about fifty percent is reliable while the remainder, based largely on secondary and often dubious sources, could be consigned to the fiction department."

Understandable, perhaps, for as the hard-bitten newspaper editor likes to observe, only trouble is interesting, and there is nothing like scandal to engage the attentions of an audience. Coupled with the temptation to judge the actions of those who lived a century ago through the lens of present-day political correctness, history often becomes subject to revision. It all became a bit too much for Ms. Mulholland, however, when she picked up an issue of the *New York Times* in 1991 to read an article citing *Chinatown* as a historical primer for understanding the practice of municipal predation upon water resources, resources to which, the *Times* writer maintained, the city had no right.

Galled that a crime melodrama, however artful, "had come to be regarded by the uninformed as a kind of documentary work on the history of Los Angeles" and dissatisfied by the paeans to Mulholland penned earlier in the century, she was moved to write *William Mulholland and the Rise of Los Angeles* (2000), a biography of her grandfather that is a meticulous birth-to-death record of a remarkable life.

While that publication was both well received and thorough, in its original form it was nearly twice as lengthy, and in the editing process certain more personal gems were unfortunately lost. In that first manuscript, Ms. Mulholland traces at some length her own history and growing fascination with her subject, including recollections of "the occasional encounters with those who simply had a friendly curiosity about my relationship to the engineer and who sometimes spoke of their admiration for 'your grandfather, who did so much for Los Angeles.'"

But in time, the burden of her family's name became increasingly apparent: "As I grew older . . . certain darker brushes not only disturbed—they paralyzed me as I had no sure defenses against them. They could come unexpectedly, as on a night in the 1940s, when I sat listening to Art Tatum's virtuoso keyboard improvisations at the Streets of Paris in Hollywood. During intermission, I fell into talk with the man next to me at the piano bar. Although in his cups, he was knowledgeable about the music and so when in the course of a genial exchange, he asked me my name, I was not prepared for his reaction.

"'You're not related to that son of a bitch, are you?' I tried to toss it off. 'Which son of a bitch did you have in mind?' But I already knew what was coming, and it did: "The one who stole the water.'"

There were other such encounters, including a moment when she took her place alongside a fellow student in an Old English Philology seminar at Cal-Berkeley, where she had relished the anonymity of being a Mulholland in a distant land. After the professor had called the roll, her classmate, who would eventually become known as the accomplished poet Jack Spicer, leaned over to whisper, "I've always wanted to meet one of *you*." Spicer's intonation left no doubt of his attitude, and the two would spend the rest of their graduate school days arguing their respective positions.

Ultimately, Ms. Mulholland says, a woman asked her how she thought her life had been affected by having been a granddaughter of William Mulholland, and she tossed off an answer: "I told her it would take a whole book to answer that." The book that ensued was in fact more about her grandfather than herself, but given the nature of the man and his work, that was to be expected.

Yet for all that has been written and debated, it seems that a certain essential story remains to be laid out, one with a clear focus on a much larger than life individual taking on an engineering project that most thought impossible, contending with forces that make our own day-to-day struggles mundane. And has it been mentioned, as they like to ask in that story dome of Hollywood, that the stakes are quite high?

Justin Kaplan, the author of *Mr. Clemens and Mark Twain*, winner of the Pulitzer Prize for biography, once said that his approach to the form was to emulate "the imaginative world of the great 19th-century novels . . . *Madame Bovary, David Copperfield* and *War and Peace* . . . [which] render individual character in the round, depict its formation and peculiarities and tell a generously contexted story with memorable scenes, a beginning, a middle and an end." I will claim no comparison to works such as those, but I will contend that in the character of William Mulholland and the real-life story that he lived is everything any novelist desires, and then some.

The truth is that without William Mulholland there might never have become a place named Hollywood, or a film industry within that place, or a carpet of San Fernando Valley lights to stare out over from a road along the mountaintops at night, and that is just a small part of what he made possible. Before Mulholland, there was next to nothing in the basins that hold 10 million or so people today, and there seemed little chance that there ever would be anything much until he went to work.

How could that be? a person might wonder. How could one man have made such a place as Los Angeles possible? Those were the questions that sent this writer once again leapfrogging back to the past.

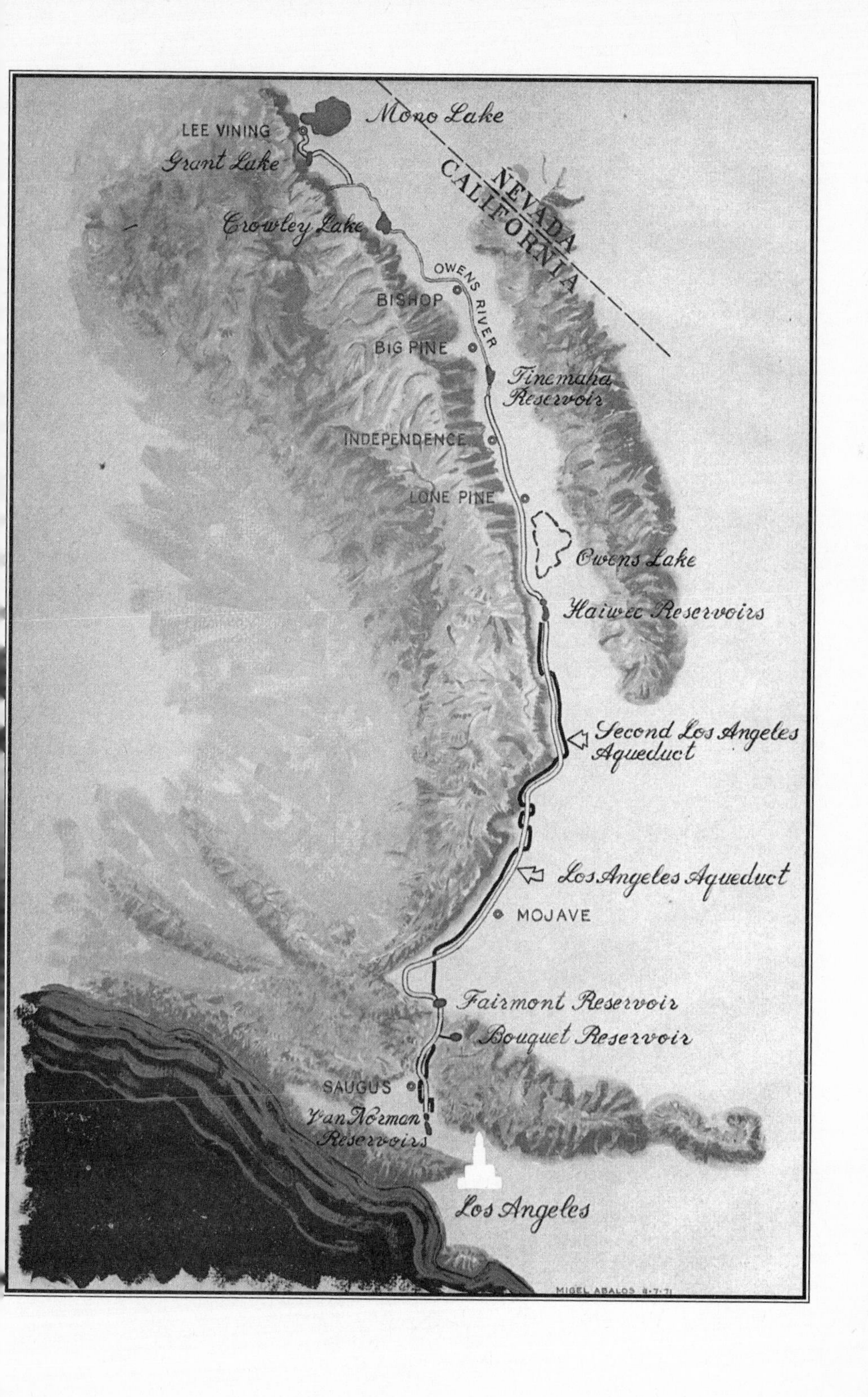
Mono Lake
LEE VINING
Grant Lake
NEVADA
CALIFORNIA
Crowley Lake
OWENS RIVER
BISHOP
BIG PINE
Tinemaha Reservoir
INDEPENDENCE
LONE PINE
Owens Lake
Haiwee Reservoirs
Second Los Angeles Aqueduct
Los Angeles Aqueduct
MOJAVE
Fairmont Reservoir
Bouquet Reservoir
SAUGUS
Van Norman Reservoirs
Los Angeles
MIGEL ABALOS 8-7-71

HOW DREAMS MIGHT END

SHORTLY BEFORE MIDNIGHT ON MARCH 12, 1928, CARpenter Ace Hopewell piloted his motorcycle up the twisting San Francisquito Canyon Road north of Saugus, about fifty miles north of Los Angeles. Through the scrub on his left, he had a moment's view of the St. Francis Dam, a looming 700-foot-wide concrete monolith, then he was into a curve and all he had was the roadway in his headlamp. He came out of the curve into a straightaway where he ordinarily would have opened the throttle, but he felt a sudden shaking—perhaps something going wrong with his engine—and instead he slowed. He was living in a construction camp next to Los Angeles Water Bureau Power Plant #1, just a few minutes' ride ahead, and there was no hurry. It was a typically cool but clear mountain night in Southern California—maybe it was a good time for a smoke.

Hopewell eased the bike off the roadway at a turnout and let his engine idle. The motor seemed steady and the shaking had stopped, but he thought he heard some crashing sounds in the distance. The spot, several miles up a wilderness road from where Magic Mountain now sprawls alongside I-5, would ordinarily be quiet enough, even on an evening in the twenty-first century. On that night in 1928, when virtually nothing existed in those reaches of the Santa Clarita Valley, his engine would have been all he heard.

Hopewell had scarcely gotten his cigarette going when a more menacing sound caught his attention. The rumble, low and rising up from the valley behind him, was a little like thunder, but that was a rare occurrence for these parts, and the crystalline sky concurred. More like a cascade of boulders down a mountainside, Hopewell thought—landslides were common in the area. He took another glance in the direction of the new St. Francis Dam that he'd passed a mile or so back, ground out his cigarette, and revved his engine. Eleven fifty-eight on a Monday night. Time to get on home, get some sleep, be ready for the next day's work.

He had no idea how drastically his "work" was about to change.

ENGINEERS AT POWER PLANT #1 realized that something was wrong when their instruments registered a sizable "bump in the line," as one put it. At the Edison Electric Powerhouse in Lancaster, operators were similarly concerned when their own lights began to flicker wildly.

Down at the St. Francis Reservoir, however, dam keeper Tony Harnischfeger's concerns had been building for several days. The dam had been completed two years before, in March 1926, and water diverted for storage there from the controversial Los Angeles Aqueduct—as "impossible" a building project as the Oversea Railway to Key West before it—had been piling up behind the walls ever since.

Only five days before, on March 7, legendary Los Angeles Water and Power director William Mulholland had finally ordered the impoundment to cease. There were now 12.5 billion gallons of water held back by the 195-foot-high dam, a goodly portion of a year's supply for the City of Los Angeles, "sufficient," as George Newhall, president of a San Fernando Valley farming company put it, "to cover sixty square miles of land with water one foot deep." One could also think of it as a section of a river ten feet deep, one mile wide, and six miles long, Newhall said.

However one envisioned it, there was quite a mass of water being stanched by the St. Francis Dam, and that was just fine by William Mulholland. The long-time, pulled-up-by-his-own-bootstraps director of the water department was often referred to as the father of the city, credited with making the modern metropolis possible when he built the politically divisive 233-mile-long Los Angeles Aqueduct between 1907 and 1913.

The acquisition of the rights to the water that now flowed to the City of Angels from a distant river on the eastern flank of the Sierra Nevada Mountains began an engineering project that ranked with the building of the Panama Canal in scope and challenge. And the fact that Mulholland, who'd never so much as finished high school, let alone set foot in an engineering class, had designed and ramrodded the project to completion, on schedule and under budget, was considered nothing short of amazing.

As if the unprecedented—and sometimes deadly—challenges of the work were not enough, the very process of acquiring the rights to the water and the rights-of-way for the passage of the aqueduct itself divided California's citizenry as nothing ever had before. The "Rape of the Owens Valley," as the water's acquisition was sometimes called by the project's critics (that phrase was first used as a chapter heading in a 1933 history entitled *Los Angeles* by Morrow Mayo), not only strained relations between Northern and Southern California interests, but was enough to draw trust-busting environmental champion President Theodore Roosevelt into the fray on the city's and "the Chief's" behalf. But all that was, in Mulholland's mind, ancient history. Recently, he had been concerned with building a series of reservoirs such as the St. Francis where more than enough water could be stored for his city should the aqueduct's delivery be threatened by extreme drought, or damage wrought by earthquake or by acts of sabotage that had been directed at the project on many occasions.

Yet Mulholland's satisfaction with the St. Francis Dam, the second largest in the system, was not mirrored by dam keeper Harnischfeger. From the very day that impoundment was halted, with waters lapping just three inches below the spillway, Harnischfeger had discovered worrisome cracks and leaks in the structure. Though he reported his concerns to Mulholland, the Chief was confident that such cracks and leaks were part of the normal settlement process for such a sizable concrete dam. Still, over the ensuing days, passersby reported that the roadbed on the adjacent San Francisquito Canyon Road seemed to be sagging in places. One motorist noted that there was water running in the normally dry creek bed below the dam, even though the dam's spillways were closed.

On March 12, only hours before carpenter Hopewell would stop for his cigarette, a troubled Harnischfeger rose early and began another round of inspections. He might have been content to live with his chief's insistence that every seep that he'd reported to date was part of a normal settling-in process for a new dam, but what he found himself staring at on this morning brought fresh concern. It was not just water oozing from a freshly discovered crack near the bottom of the dam, it was *brown* water, which suggested to Harnischfeger that the water had begun to erode the foundation of the dam itself. The dam keeper got on the phone and insisted that the Chief come out and see for himself.

At about 10:30 in the morning, Mulholland arrived from Los Angeles, along with his chief assistant, Harvey Van Norman. A worried Harnischfeger escorted them on an inspection tour, sure that the two would appreciate his concerns. But in the end, Mulholland shook his head. There was simply no cause for alarm. Everything they had seen was to be expected. Cracks were common in a concrete dam of this size. And the muddy color of the water running down to the creek bed was caused by runoff from a recently constructed access road, Mulholland said, pointing to a gash in the nearby canyonside. Harnischfeger should keep his eyes peeled and report if anything extraordinary turned up, but meantime he was to rest assured. In William Mulholland's opinion—and there was absolutely no authority in Southern California more highly respected in such matters—the St. Francis Dam was safe.

DESPITE MULHOLLAND'S CERTAINTY and the respect he was accorded within his profession, it is an open question how much reassurance Harnischfeger took from the Chief's assessment. After all, Harnischfeger lived with a woman named Leona Johnson and his six-year-old son, Coder, in a small cottage on the floor of the San Francisquito Canyon, about a quarter of a mile directly downstream from the dam. As a matter of fact, a motorist driving the canyon road about 11:30 that night reported seeing a light in the canyon near the foot of the dam, suggesting that Harnischfeger may have been down there poking about with a lantern at about the time that

Ace Hopewell heard that odd sound of boulders crashing down a mountainside.

What can be known for certain is that Hopewell—the last man alive to have seen the structure whole—had actually heard the total collapse of the St. Francis Dam. It is also known that the waters exploding down the canyon were 140 feet high when they pulverized the cottage where Harnischfeger and his family lived and, seconds later, the bully brick and concrete edifice that was DWP Power Plant #2. The fully clothed body of Leona Johnson was later found wedged between two blocks of concrete swept down from the broken base of the dam. Neither Harnischfeger's body nor that of his son was ever found.

Catastrophe would multiply, the wall of water catapulting down the ordinarily dry bed of the Santa Clara River, scouring a path a mile and a half wide all the way to the Pacific Ocean, fifty-four miles away. Eighty-year-old C. H. Hunick told one rescue worker at a hastily erected field hospital near Saugus that he had lived in a ranch house about a mile and a half below the dam. "When the water hit the house, it folded like it was built of cards," he said. Hunick grabbed onto a chunk of wood and floated for miles, caught in the roaring current. Nearing exhaustion and about to lose his grip on what he realized was a piece of his home's roof, Hunick felt a hand grab his arm in the darkness.

"Is it you, Dad?" It was the voice of one of his sons, come miraculously from the darkness.

Hunick described how his son hauled him to a plank he'd been using as a life raft. The two floated on together until the elder Hunick lost consciousness. He awakened in the hospital, and from an attendant wanted only to know where his two sons were. The worker stared and shook his head. The bodies of Hunick's sons lay in a temporary morgue nearby.

~~~

WORKERS AT POWER PLANT #2, a little more than a mile downstream from the dam, had only a moment's glimpse of that avalanche of water rushing toward them before they and their plant and their families living in cottages nearby were swept away, all but three of them drowned. More than a hundred men working on a construction project for Southern California Edison were camped in tents near the mouth of San Francisquito Canyon when the waters hurtled out from between the canyon walls only minutes after the dam had burst. Eighty-seven of them drowned. Many who lived to describe the experience were those who instinctively buttoned the flaps of their tents as the enormous wave rushed down. They found themselves floating to the top of the swirling waters as if they were riding in huge canvas balloons.

The flood took out every road and bridge between the dam and the coast, including the Southern Pacific Railway line connecting Oxnard and Ventura and the freight line between Saugus and Montalvo. In all, the dam's failure took at least 450 lives, a disaster outdone in California history only by the 1906 San Francisco earthquake and fire. Thousands of homes were destroyed, and damage to property would make it the greatest civil engineering calamity of the twentieth century.

In the aftermath, grief and outrage were the order of the day. A coroner's inquiry was convened to uncover the cause of the disaster, and the legendary Mulholland, at seventy-two, faced a firestorm of criticism and scrutiny that neither he nor anyone else could have conceived. To modern readers conditioned by the duck-and-cover responses from public officials following any disaster, Mulholland's response to the onslaught might seem as noteworthy as his stout accomplishment in building the aqueduct in the decades before.
~~~

"Don't blame anyone else, you just fasten it on me," he said. "If there was an error in human judgment, I was the human, and I won't try to fasten it on anyone else." Devastated by the event that refashioned him from civic hero to villain in an eye-blink, Mulholland would at one point confide to a reporter, "I envy those who were killed."

DISTANCE BETWEEN TWO POINTS

ACCORDING TO CALIFORNIA DEPARTMENT OF TRANSportation figures, upward of 275,000 vehicles travel Interstate 5 through the Newhall Pass dividing the Santa Clarita and San Fernando watersheds each day. There is no way to tally the total number of individuals inside those vehicles, but taking into account the tendency of the average American driver, it is safe to speculate that at least 275,001 travelers per day have the opportunity to glance eastward of the thundering highway near its LA-side crest and behold "The Cascade," as William Mulholland termed it, the concrete spillway marking the termination of the Los Angeles Aqueduct, the place where the waters drawn from the Owens Valley enter the San Fernando Valley.

While some likely know what they are looking at, it is probably just as fair to say that few of these quarter-million-plus individuals are much interested, being more concerned with overheating engines, lurking California Highway Patrol officers, traffic jams, and troubles looming at the end of a formidable commute to work or back home. Certainly, few average daily drivers or passengers could fathom the enthusiasm of the 30,000 (or 40,000 to 50,000, depending on whose figures one uses) who swarmed the nearby rugged hills back on November 5, 1913, to watch the water that would make Los Angeles as we know it finally tumble down. These days, the Newhall Cascade might strike some as a curiosity, especially on the occasions when it is carrying water (the water is often coursing through the huge steel pipes or "penstocks" adjacent), but even for those who notice or know what they are looking at, the crashing water is likely a given, as is the sprawl of 10,000,000 or so people that has come to be where a century and a quarter before there were scarcely 50,000.

It would take a book of its own to describe what *wasn't* to be found in Los Angeles at the turn of the twentieth century. Only due to the fact that some contiguous lands remain in the public domain and others are so sheer and forbidding as to deter even the most resolute developer can current-day residents appreciate the tough desertscape that pioneers in the region had to contend with back then.

It is difficult to imagine a dusty Los Angeles basin where virtually all the homes and businesses hugged the course of a feckless stream draining the forlorn San Fernando Valley from west to east before curving around the tail of the Santa Monica Mountains between today's Griffith Park on the west and Glendale to the east. From that point, it is a little more than thirty miles southward to the emptying of the Los Angeles River into the Pacific, and for most of the time from the city's founding by the Spanish in 1781, its days as a part of independent Mexico (from 1821 to 1848), and its early days as an incorporated city (its formation on April 4, 1850, actually predated by five months the designation of California as a part of the United States), almost no one lived any farther from the river (it was originally called the Porciuncula) than a gravity-fed irrigation ditch or a horse-drawn water cart or a few miles of leaking wooden pipes could reach.

In contrast, settlement of the founding colonies of the East Coast was often a matter of finding a place where floods and tides would not drive a family from its home. Water was everywhere in abundance on the opposite coast, and often too much so. In fact, its staff-of-life properties aside, the most important function of water in the early days of the Union was as a means of transportation. At the time of the Revolutionary War, every Colonial settlement of any significance was situated upon a navigable body of water.

In the arid West, however, water was not so much an "aid" to civilization as a sine qua non. Settlers didn't go much of anywhere *on* the Los Angeles River—they simply didn't go far from it. The history of development in the American West is, as any number of authoritative works have shown, largely intertwined with the ability to find, develop, and maintain a reliable source of water. Accordingly, for well beyond the first hundred years of its existence, the likelihood that Los Angeles would ever become a major city was very much in doubt.

By 1890, and given the limitations of hydrological practice, it was clear that Los Angeles had tapped its "mother ditch," as the Spanish referred to the main irrigation canal virtually synonymous with the river, just about dry. While there was a general understanding that much of the water of the Los Angeles River Basin ran its course well beneath what was visible at the surface (William Mulholland liked to call it an "upside down river"), no sophisticated equipment existed that allowed for an accurate mapping or precise measurement of underground reserves. Attempts to dig wells in one area often resulted in parching those previously dug elsewhere. In addition, while the interests of ranchers and farmers in irrigating relatively vast expanses of land were real, the implications of a water shortage for a city's domestic consumption were, in terms of numbers and politics, even greater.

It seemed clear to parties interested in governing a city, as well as to those boosting its efforts to grow, that while the current level of agricultural activity and a minimal amount of dry ranching could continue, and that a population of 100,000 or possibly 200,000 might be maintained as well, the city's water source was just about tapped out. There were fruitless water-seeking forays into the nearby San Gabriel Mountains and claims from speculators and landholders surrounding desultory streams that their holdings could be acquired to solve the city's problem, but the so-called solutions were stopgaps at best. All those acres of undeveloped land destined to sit idle evermore; the surrounding settlements such as Pasadena, begun with promise but soon to die of thirst; all those imagined cities existing only as glints in promoters' eyes—including altitudinous Hollywood—sure to die aborning, unless water could somehow be found.

It was out of such desperation that a 250-mile journey from Los Angeles to the far-flung Owens Valley—a mythic cradle of waters rumored to exist in the distant Sierra Nevada Range—took place back in 1904. Taking that same trip 110 years later has little to do with desperation, or myth, and it is certainly a quicker trip than one made by horse and buckboard. Still, in retracing the journey, the modern traveler gains an inkling of how unlikely was the connection between two such disparate physical points.

IT'S DIFFICULT INDEED TO conceive of such a connection while, I-5 traffic willing, a twenty-first-century driver soars over the crest of the Newhall Pass and descends into the Santa Clarita Valley. While that valley actually lies within the borders of Los Angeles County, it is something of a world apart. It is within the realm of reason to commute to Los Angeles from the far-north San Fernando Valley communities of Chatsworth or Northridge, and a resident of Reseda or Canoga Park can vaguely be construed to be a member of the Angeleno fraternity, but by the time one makes it the twenty-six miles or so from the site of the Mulholland Memorial Fountain at the intersection of Los Feliz Boulevard and Riverside Drive to Valencia or Saugus in the Santa Clarita Valley, the concept of the city has begun to fade. For most Angelenos, the most identifiable feature of the Santa Clarita Valley—save for the few who might have forayed out to play a spectacular Valencia golf course since gone private—is likely the sprawling amusement park known today as Six Flags Magic Mountain, a 262-acre roller-coaster-heavy theme park just off the freeway in Valencia, thirty-five miles north of the Los Angeles city center. (Magic Mountain, which opened in 1971, attracts about 2.5 million visitors each year; Disneyland, situated about an equal distance south of Los Angeles, opened in 1955 and hosts 16 million or more each year.)

On a clear, cold Thursday afternoon in January, however, there is no typical clot of traffic at the Magic Mountain turnoff, and it is only a minute or two to the next interchange and ten more or so of largely unimpeded twisting and turning beneath the pines and eucalyptus through an *ET*-worthy Santa Clarita suburb-scape to a turnoff for San Francisquito Canyon Road. If the connection to Los Angeles had previously seemed tenuous, at this junction it frays altogether.

The two-lane blacktop road ahead, its course slightly modified in the eighty or more years since Ace Hopewell motorcycled it, winds northward along a mostly treeless valley floor for a few miles, passing a series of low-slung ranch homes of a style unchanged since the '60s, a number of them hard by the trail, others surrounded by pastures and horse farms. Every so often a cluster of mailboxes appears atop a length of whitewashed two-by-six, suggesting any number of homes somewhere out in the flats that were once scoured clean. There is not a lot of traffic here—the occasional working pickup, a few mom-vans, once in a while a throbbing lowrider—and the smogless vista capped by a cloudless blue sky suggests a high desert scene from just about anywhere in the limitless stretches of the American frontier west of the Pecos and north of the Rio Grande. Everything in these parts seems to be waiting, waiting, waiting.

There are only three or four miles of habitable land to pass through before the canyon walls narrow quickly, and anyone who's read of the 1928 disaster gets a flash of what the workers in the Edison Camp felt when they saw a wall of water hurtling out from the looming jaws of rock nearby. A couple more quick turns, and the canyon has narrowed from a mile wide to a half mile and then to a hundred feet or so—the road has become a twisting track—and at the point about six miles up the canyon road where Power Plant #2 sits, the walls are so sheer that even an antelope would have been out of luck when the flood pounded down (press accounts told of a single power-plant worker who managed to claw his way up the cliff side of the water). Though there is a turnout at the stolid power plant rebuilt in the 1930s, and no shortage of historical markers to read there, it is not a place for the claustrophobic or the suggestible to linger.

It is a mile or so on up canyon to the place where the St. Francis Dam once blocked the water's passage to the chute below, but a seeker has to be looking hard for what remains of that structure. Though the road once directly skirted the remains of the ruined dam, the so-called Copper Fire of 2002 resulted in a realignment of the route, and only the resolute will spot the place to pull off and walk back down an abandoned stretch of blacktop to the place where William Mulholland once stood with Harvey Van Norman to reassure dam keeper Harnischfeger that the structure was safe. It is said that, in the wake of the Chief's departure back to Los Angeles that day, a group of young electrical engineers took their lunch atop the broad concrete curve of the dam. What the Chief knew was all that needed to be known. There is no marker at this site.

Still, it is possible to climb a few hundred yards or so up a rugged talus slope and stand atop the remnants of the dam's wings and look northward toward the place where Power Plant #1 still sits, though there is no longer a three-mile-long, 12-billion-gallon impoundment of water over which to gaze. Water still courses the route, but it is unseen these days, all of it ensconced within huge pipes, not only slaking thirst and filling tubs but churning turbines on the way. What *is* visible is a broad, uninhabited swath of scrub-covered Angeles National Forest, ringed by mountains, where a hundred years ago, a man supposed that a displaced river might run.

IT IS A BIT LESS than twenty miles on through the forest, through Green Valley and across the canyon crest of the Sierra Madre to Lake Elizabeth, which is often mistaken for a reservoir instead of the naturally occurring sag pond that it is, a feature that in fact constituted a formidable obstacle to the long-ago plans of William Mulholland. Lake Elizabeth and environs constitute a pleasant enough area these days, with RV parks, recreational opportunity, and—judging by the number of billboards—home-site and development acreage abounding, but any vestiges of urban living by this point lie far behind. Beyond the ridgeline to the north of Lake Elizabeth, all roads lead down toward the arid plain of the Antelope Valley, where the namesake pronghorn once thrived and where, at another time of year, one could be diverted by a visit to the nearby California Poppy Reservation and a stroll through a nearly 2,000-acre carpet of the eponymous orange flowers.

In January, however, poppies are just a fever dream for a desert traveler, as is the thought of the buried aqueduct that parallels the narrow blue highway route northward toward civilization's last outpost—in Mulholland's time, as now—at Mojave, about a hundred miles north of Los Angeles. Though the reasons to travel to Mojave are presently largely practical, the point of this journey is anything but. Still, if one were compelled for some reason to follow the route of Mulholland's aqueduct, one would find a way downhill and eastward across the seemingly limitless valley to California's Route 14, which leads to Mojave, which, whatever one might be moved to say about it today, was not very much at the turn of the twentieth century.

Sprouted in 1876 from the high desert plains (its altitude is 2,762 feet, and its average annual rainfall does not fill three-quarters of a cup), Mojave was originally a work camp for the Southern Pacific Railroad and later a terminus for the fabled twenty-mule-team wagons bearing borax mined in Death Valley on the Nevada border, 200 miles to the northeast. Though it would later become something of an aerospace and military aviation center (Edwards Air Force Base and China Lake Naval Air Station are close by), Mojave's chief appeal to Mulholland was simple: in 1907, you could get there from Los Angeles by train, and at the time, that was as close to the distant Owens Valley as one could easily travel.

From Mojave all the way to Bishop, the northernmost settlement in the eighty-mile-long Owens Valley, it is a little more than 170 miles, a journey that took Mulholland and his original traveling companion to the area, Fred Eaton, a former mayor of Los Angeles, several days. A century or so later, it takes the undiverted traveler a little less than three hours to make the trip, with hardly a traffic signal on the way. Most who travel the route northward (California 14 merges with US 395 about forty miles north of Mojave, near China Lake) are indeed undiverted, bound for the jump-off to the Mount Whitney Trail or the ski or summer resort areas near Mammoth Lakes, another forty-five miles beyond Bishop in the High Sierra. Most of those travelers likely view those intervening miles as an ordeal to be endured, more or less the way a child measures the run-up of days from Thanksgiving to Christmas.

It is a generally gradual climb out of Mojave for the eighty or more miles to the Haiwee Pass, where one gains the Owens Valley, and there is very little along the way to suggest that an immense amount of water is thundering in the opposite direction, closely parallel to the driving route. To the west, the southern terminus of the Sierra Nevada looms treeless and foreboding, and to the east, the view over China Lake and beyond suggests that Death Valley is a forgiving description for this terrain. No towering eucalyptus here, no pines, no grassy median strip or shoulders. Just sand and scrub and a landscape as riven as a hag's face.

Still, about twenty miles north of Mojave, one can turn westward off Route 14 and twist along an unremarkable desert blacktop road for a mile or so to where a startling sight appears: in the midst of country as dramatic as any from a Technicolor Western of the 1950s, a leviathan of pipeline suddenly heaves into view. It is the massive Jawbone Siphon, 8,000 feet of one-and-one-eighth-inch-thick steel, ten feet in diameter, plunging downward from a ridge to the north, then hurtling a mile or more across the rugged valley floor, then charging back up the canyon wall to the south, plug-full of water rushing toward Los Angeles.

As remarkable as the mechanical acrobatics is the visual contrast: 80-million-year-old late-Paleozoic terrain traversed by a steel-encased river dreamed up a century ago on its way to a city of vast importance—for a relative instant in geological time, at least. Jawbone Canyon is not the only place where the modern and the ancient sit side by side, but there are few where the antipodes are so starkly laid out.

Edward Abbey, the highly regarded western writer of *Desert Solitaire* (1968), was once given what he thought of as a plum assignment—writing the text of a Sierra Club publication touting the beauties of an area in the Appalachian Mountains. When he returned from his initial research trip, this writer asked him how it went. "I had a really hard time there," Abbey said. Familiar with his prickly personality, I asked if it had to do with the people who were showing him around. "Oh no," Abbey said. "They were fine. It was the place itself. Just too damn much to look at." In the elemental landscape of Jawbone Canyon, no such problem presents itself.

There are other spots of interest along the climb toward the Owens Valley, of course. A few miles up the road from Jawbone is Red Rock Canyon State Park, where the rugged cliffs have provided a striking backdrop for films as diverse as *The Mummy, The Big Country,* and *Jurassic Park,* and farther along, in the fold just below the pass that leads up into the valley, one can amble a few miles eastward down an access road for a glimpse of the seven miles or so of reservoirs created by the North and South Haiwee Dams. It is here that the waters of the Owens River have their last wash of sunlight before entering the pipes and concrete-covered conduits that take them on to the San Fernando Valley, and the view provides another of those striking contrasts common to the territory: the stark peaks of the Coso Range to the east, the snowcaps of the Sierra Nevada looming to the west, and a seemingly impossible expanse of blue water stretched between.

From Haiwee, it is only a skip and a little jump over the pass to a wide spot on Highway 395 that is Olancha, home to a couple hundred hardy souls, where one enters the Owens Valley at last, nearly 200 miles north of Los Angeles. Olancha marks what was once the original end of the line for the Owens River, which descended through the valley for some ninety miles to this place where the Cosos butt up against the Alabama Hills trailing from the Sierra Nevada, forming a natural impasse for the river. There were some few times in the late Pleistocene when extraordinary runoff from the Sierra swelled the waters of Owens Lake into overflow down the canyons past Haiwee (think: how China "Lake" got its name). But Owens Lake was for most of its last thousand years a trapped body of water, ever increasing in its salinity, about twelve miles long and eight miles wide, ranging in depth from thirty to fifty feet. In its heyday, the lake was a significant nesting and migratory stop for ducks, geese, grebes, gulls, sandpipers, and various other species. Today, and despite the expenditure of about $1.2 billion by the City of Los Angeles to regenerate its marshlands, it is virtually dry, a vast alkali flat that is the source of some of the state's most significant air pollution when seasonal winds kick up.

It should also be mentioned that the traveler has been resident for some thirty miles now in the County of Inyo, the diversified topography of which, as its chief chronicler, W. A. Chalfant, once observed in *The Story of Inyo* (1922, 1933), "is matched by no other on earth." Early on in his history, Chalfant points out that nearby Mount Whitney, the highest peak in the continental United States at 14,500 feet, is not all that formidable when compared to some in the Himalayas. But, as he notes, this county is also home to the lowest point in the United States, in Death Valley, at nearly 300 feet below sea level. Furthermore, with only about eighty-five miles separating them, one is clearly visible from the other.

Even without a side trip to Death Valley or a surveyor's instruments, any late-afternoon traveler to the Owens Valley soon recognizes the Inyo extremes Chalfant refers to. One moment, it seems, the land laid out north of Olancha and nearby Cartago is awash in sunlight, and in the next, all is shade. If a glance at the 12,000- to 14,000-foot Sierras looming to the west were not enough to explain, a stop at the Interagency Visitor Center at Lone Pine, another twenty miles along will clarify: among other things, a huge topographical model of the region shows that the rate of decline from the top of the nearby Sierras, where the southernmost glacier in the United States is found, to the 4,000-foot floor of the Owens Valley is among the steepest in the world. While the western slopes of the Sierras rise up at a relatively leisurely rate, taking some fifty miles to ascend from the San Joaquin Valley, it is only ten to fifteen miles eastward from the tips of Mount Whitney and its cousins to the riverbed that courses the Owens Valley floor. This precipitous drop makes for a relatively early twilight in all seasons, though that is hardly the most significant effect.

Moisture condenses evenly and reliably as the winds off the Pacific rise up those western slopes. Snowpack on Mount Whitney and the Mammoth Lakes area some eighty miles to the north can reach seventy feet or more in winter. But once those winds have cleared the Sierra summits, there is precious little moisture left to drop. The average rainfall in the Owens Valley itself is about six inches yearly, not much more than falls on relatively featureless Mojave. The California Department of Water Resources description of the valley's groundwater basin offers statistical confirmation of what is apparent to any visitor's eye: of the basin's 1,000 or so square miles of traversable territory, 1 percent is urban in nature; another 5 percent is taken up by agricultural endeavors; the remaining 94 percent is characterized as "natural."

Some more attuned to city dwelling might be inclined to substitute "forlorn" for "natural," and there is surely something of *The Last Picture Show* or *High Plains Drifter* to be encountered in the valley towns of Lone Pine (pop. 2,035), Independence (the Inyo County seat, with 669), Big Pine (1,730), and Bishop, which is, with its 4,000 or so residents, the only actual city in Inyo County. Temperatures here range from winter highs in the 50s to nearly 100 degrees Fahrenheit in the summer. There are no roads that traverse the formidable Sierra to the west, and while the highway north offers access to Yosemite and Lake Tahoe, most travelers from the southland think of the route as the one that ends at Mammoth. Though the rugged White Mountains to the east can be skirted by car, those roads lead only to the true *terra incognita* that lies between Las Vegas and Reno. Or, to put it another way, the Owens Valley lies pretty much at the end of the road. Most people get out the same way they come in.

And still, it is about as dramatic a landscape as can be imagined, a singular apparition that transcends any single evaluation of it. As Chalfant puts it, "Nature has written here, in bold strokes, studies more fascinating than the little affairs of humanity." Some first-time visitors may experience a certain claustrophobic effect when that early twilight suddenly sweeps down from the Sierra, but the feeling goes beyond the simple physical fact of being caught in a dozen-mile-wide graben between two towering parallel mountain ranges. (Geologists suspect that at the time the Sierra Madre and the White Mountains were formed, the intervening valley floor was as much as 10,000 feet deeper—over the 80 million years or so since, glacial plowing and erosion off the peaks have filled the valley to its current level.)

This is a place where, simply put, the geologic immensity of the earth asserts itself. There's a bit of this drama at Jawbone Canyon, but here the indifference of the universe is both more dramatic and more sublime, leavened by the drama of the surrounding snow-capped peaks, the dome of an ordinarily crystalline sky, and the vista of the green-gold plains below. It is permissible, even proper, to feel dwarfed.

There is civilization in the Owens Valley, of course. There is a sizable Crystal Geyser Natural Alpine Spring Water bottling plant in Olancha, and in Cartago there are reminders of that town's late nineteenth-century prominence as a steamboat port where bullion mined in the Inyo Mountains across then wave-capped Owens Lake was off-loaded and hauled by mule team to Los Angeles. Not far north of Cartago stands another testament to early valley enterprise, the Ozymandian remains of a Pittsburgh Plate Glass factory put in place to utilize the mineral compounds deposited over the centuries in the lake bed. The plant was abandoned in the 1960s but still stands intact, looming above the nearby dunes like some impossible mirage from the Rust Belt.

Lone Pine, the town from which the main trail to Mount Whitney is accessed, has a high school and a hospital, and is home to the Lone Pine Indian Reservation where 200 or so members of the Paiute-Shoshone tribe, original settlers of the valley, still live. Also near Lone Pine is the national historic site memorializing the Manzanar War Relocation Center, one of the ten Japanese-American internment camps where, during World War II, Japanese-American citizens were displaced by the US government. Any number of films have been shot in the picturesque Alabama Hills nearby, including the Humphrey Bogart classic *High Sierra* and *Bad Day at Black Rock,* which features the memorable appearance of a one-armed, black-suited Spencer Tracy come to deliver justice to the West well in advance of the High Plains Drifter.

Sixteen miles farther north of Lone Pine is the Inyo county seat, Independence, where three blocks west of the rustic courthouse sits the Eastern California Museum, housing an Aladdin-esque trove of documents, artifacts, and exhibits pertaining to the natural and cultural history of the region, including its mining heyday, its Native American culture, the camp at Manzanar, and, of course, the building of the Los Angeles Aqueduct. The museum's grounds also feature a native plant garden and an exhibit of various agricultural tools and implements used by settlers trying to carve a space in a difficult land. On the day of this inspection, there was a hawk perched on the museum's gabled roof, as patient and untroubled by visitors as a wind vane, and while no rattlers coiled and buzzed in the nearby grounds, it would have seemed appropriate if they had.

It is about fifteen miles from Independence northward to Big Pine, a town of nearly 2,000, where the Paiute-Shoshone tribe keep their offices, and from which one can make a side trip into the White Mountains for a look at the oldest living organisms on Earth, the ancient bristlecone pines, one of which is thought to have germinated in 3050 BC. From Big Pine it is another fifteen miles on to Bishop, where the road widens and traffic lights sprout, and national franchise stores butt up against cowboy-themed mom-and-pop restaurants, bars, and dry goods purveyors.

The Bishop Paiute tribe has a reservation just north of town, and the offices of the Inyo National Forest, which encompasses vast amounts of land but very few trees, are also located in Bishop. Bishop Creek, which gives the city its name, burbles beneath US 395, here to join with the Owens River, and there is at least one pleasant inn astride its banks. There is also a local steakhouse where the meat is well marbled, the drinks are heavy on the pour, and you might catch sight of a rancher in a starched white shirt with a curl-coiffed lady on his arm, come in for a night on the town. Bishop is the last natural stopping point before one climbs northward out of the valley to ski, or hike, or find a way on to Yosemite or Lake Tahoe and the world of institutionalized tourism. It is as good a place as any for a pause.

At Bishop, the traveler is about 265 miles north of Los Angeles in highway terms, though light years away by many other measures. There are about 10,000,000 people living in Los Angeles County and fewer than 20,000 in all of Inyo. To make the transition from cosmopolitan international seaport to high desert outpost in the course of a day is not completely unlike the sort of transport dreamed up by H. G. Wells. Though this version of time travel is utterly real, it is in its own way as disorienting as that of fiction, especially given the essential and enduring connection that stretches between two such places.

A few miles back down the road, about thirty miles south of Bishop and perhaps ten north of Independence, there is a graveled turnoff that ambles eastward through uninhabited scrubland in the direction of the White Mountains. There are a couple of cattle gates to be opened (and closed behind) as one descends the gentle grade, an experience in itself for anyone whose idea of civilization's breakdown is a bottleneck at the Newhall Pass.

A couple of dusty miles along, the road opens onto a broad sandy turnaround with a few cottonwoods shading a porta-potty and a sign to remind visitors that there is to be no overnight camping in this spot. For anyone who has read of the Owens Valley Water Wars or who is aware that the City of Los Angeles to this day derives a significant portion of its water from this source, the view of the Owens River from the banks here might seem incongruous. Though a sign erected by the Los Angeles Department of Water and Power warns of swift water ahead and demands that swimmers, rafters, tube riders, and boaters exit the river at this point, "swift" is surely a relative term.

As it approaches the diversion point that marks the beginning of the 233-mile-long Los Angeles Aqueduct, the Owens River is a languid stream perhaps fifty feet wide, cutting through gently descending scrub and pastureland. Though seasonal runoff can affect the river's flow, there is scarcely a ripple—certainly no class-five rapids—here, and for those in whom the term "river" conjures up images of the Columbia, the Ohio, or the Potomac, the view could seem diminished.

And still, this is the point at which it all began, the point at which it still begins. More than a hundred years ago, a self-taught engineer and a former mayor of a desert town with far more promise than water traveled to these banks by buckboard along a route that legend has it could be traced by the whiskey bottles the pair discarded along the way.

It could be difficult for a modern-day traveler to stand on these placid banks and imagine a no-nonsense William Mulholland surveying such a scene, then turning to the storied promoter Fred Eaton to announce that they had indeed finally found the answer to the question "What will make Los Angeles possible?" But as history tells us, Mulholland was an unusually perspicacious man, and that is essentially what he did.

LUCK OF THE IRISH

THE STORY OF HOW MULHOLLAND MADE IT TO THE banks of the Owens River in the first place is very nearly the equal of what he accomplished after he made the trip, yet another of the many improbable rags-to-riches tales arising from America's Gilded Age. The period, named after a satirical work by Mark Twain, is generally thought of as spanning the last quarter of the nineteenth century, when the country emerged as a world economic and manufacturing superpower. Twain meant it as a derogatory term, but to many these days, the words conjure up a time when larger than life individuals—often of humble origins—accomplished improbable feats. There was no shortage of wretched excess in the era, and many a fool's errand was essayed, but certain of the real-life tales transcend dismissal.

Though Mulholland never amassed the personal fortune of a Carnegie or a Rockefeller, it can be argued that his legacy surpasses everything that those two left behind. Certainly, what he overcame to achieve a position of influence rivals any Horatio Alger–style narrative. Much of what is often told about William Mulholland's early life can be found in an unpublished master's thesis by Elizabeth Spriggs, "The History of the Domestic Water Supply of Los Angeles," written for the University of Southern California in 1931. Ms. Spriggs not only interviewed Mulholland's daughter Rose for the piece, but was also able to speak with Mulholland on at least two occasions. One of her most intriguing sources, however, is an autobiographical sketch written by Mulholland himself in 1930 that was in the possession of Ms. Spriggs at the time of her writing. Likely penned at Ms. Spriggs's behest, the eight-page document disappeared for more than eighty years until an employee of the Los Angeles Department of Water and Power, looking for an arcane item in the city archives, chanced across it.

The sketch, by an often guarded Mulholland, was a vexatious loss to historians, though now that it has reappeared, it seems that Ms. Spriggs made judicious use of her source, as she did of the manuscript of a Mulholland biography penned by his cousin Ella Deakers, whose family had preceded Mulholland in emigrating from Dublin to Pittsburgh in the 1860s.

Certainly, Catherine Mulholland adds a great deal in her meticulous autobiography of her grandfather published in 2000, and some of the most colorful anecdotes about his roots come from Mulholland himself, passed along by a number of associates and journalists undoubtedly in the thrall of an Irishman as wedded to the concept of a good story as to the courtroom truth. He once claimed to have cast his first vote in a failed attempt to propel Samuel J. Tilden into the White House in 1876, but if he did so, it would have been an illegal act, for Mulholland did not become a US citizen until a decade later.

While Mulholland may have embellished at times and misremembered at others, there is no doubt that he was born in Belfast, County Antrim, on September 11, 1855, the son of Hugh Mulholland, a guard for the British Mail Service, a sinecure that ensured survival if not wealth. His mother was Ellen Deakers, who had grown up (as had her husband) in Dublin, but who as a young girl spent a period of time with her family in the United States, where her father created a profitable drapery business in Pittsburgh.

Sometime in the early 1850s, when she would have been no more than eighteen, Ellen returned to Ireland to marry Hugh. She gave birth to three sons in rapid succession: Thomas in 1853, William in 1855, and Hugh in 1856. There would be two daughters as well, both of whom died of pulmonary tuberculosis, or consumption as it was called at the time. The disease would also take Ellen herself in 1862, shortly after she gave birth to a fourth son.

William Mulholland, who was seven when his mother died, would often tell outsiders, including Ms. Spriggs, that he had no recollection of his mother, but Catherine Mulholland passes along family memories of the Chief speaking candidly of his mother's lively, playful nature, a welcome tonic to his father's stolid seriousness. What can be certain is that in 1865, Mulholland's father, at thirty-six, left with four boys ranging in age from three to twelve to care for, took a second wife, Jane Smith, then thirty. In short order there were three more children born, and by 1870, William, an impatient student who had already run off to sea at the age of fourteen, signed up as an apprentice for the British Merchant Marine.

It was not owing to financial necessity or familial tension, according to Mulholland. "I had always wanted to travel," he wrote in his sketch some sixty years later, "nurturing this desire since I was ten years old." Given the first opportunity, he says, "I became an apprentice aboard the *Gleniffer* . . . without the whole-hearted approval of my father." Though he returned and gave schooling one more brief attempt, Mulholland was soon back at sea, where he would spend four years on ships that plied the Atlantic between the British Isles, the United States, and the Caribbean.

Though the life was full of the sorts of adventure that Mulholland had fancied, there was little pay involved, and it eventually occurred to Mulholland, as he put it, that the sailor's life "would get me nowhere in a material way." Thus, in June 1874, the same year that a German immigrant by the name of Levi Straus received a US patent for copper-riveted blue jeans and about the same time that a general named Custer was dispatched to the Black Hills to keep an eye on the natives there, William Mulholland stepped off a ship in New York harbor and at nineteen began a life in America.

It may not have been the most propitious time to start. While the US economy had boomed in the aftermath of the Civil War, with railroading—especially western railroading—leading the way, a downturn began in 1873 that would be known as the "Great Depression" until the even worse economic troubles of the 1930s. A quarter of the workforce in New York was off the job and once-popular President Ulysses S. Grant, just elected to a second term, faced a firestorm of criticism for his inability to right the economic ship.

Mulholland soon made his way to the Great Lakes, where his experience landed him a job aboard a freighter. When winter came, he took a job in a Michigan lumber camp, where he cut open his leg in a logging accident. The wound developed an infection, and Mulholland overheard the camp doctor mention to a colleague that the leg would probably have to come off. At his first opportunity, Mulholland struggled out of his infirmary bed and somehow made his way southward to Cincinnati. Broke and alone, seemingly out of prospects, he felt that he had reached his nadir.

"Why bother?" he said, according to a story later passed down by his daughter Rose. There seemed to stretch ahead nothing but years of soul-crushing work and miserable pay. And then, as Mulholland tells it, he found himself limping past a local church where from the doors the voices of a choir of boy sopranos issued, singing the *Gloria in Excelsis*. Mulholland wandered inside the church, where he sat for some time in meditation. Slowly, his gloom began to lift. Though he had no idea how he would make meaning of his life, he would try. Truly, he wanted to live.

In the days that followed, the infection in his leg abated and Mulholland made the acquaintance of a tinker who had been looking for a helper who wouldn't mind the traveling life. In the young Mulholland he had found his man. In fact, Mulholland once told his daughter Rose, he found the life of buckboarding from farm to farm, sharpening scissors and mending clocks, as pleasant a time as he had ever experienced. Only the sense that there was surely something more meaningful for him to do spurred him to make a change.

Mindful that his uncle had forged a successful business in Pittsburgh, Mulholland parted ways with the tinker and made his way to that city in the fall of 1875, where he was joined by his brother Hugh Patrick, who had jumped ship from the British navy shortly before. By the time the pair arrived, Richard Deakers's dry goods store was thriving and he was willing to put the two boys up and set them to work at clerking, an occupation that allowed Mulholland to develop, he was told, into "an excellent salesman with a business reputation in the county."

The arrangement lasted for something more than a year before Mulholland was again compelled to make a change. In his sketch, Mulholland says that he was inspired by reading Charles Nordhoff's *California: For Health, Pleasure, and Residence,* a rhapsodic tract that the popular writer had published in 1873. It was a book he had never heard of previously, Mulholland wrote, but it was one that "aroused such interest in me in this country [California] that I had no peace until I came here."

Family lore, however, suggests a darker cause. Sometime prior to Mulholland's arrival in Pittsburgh, Richard Deakers and his wife, Catherine, had also taken in another relative referred to only as "Uncle Hobson," a miserable wretch already emaciated by tuberculosis. By 1875, two of the Deakers children had died of the disease and two more were exhibiting symptoms. Given that three of Catherine's brothers had gone off to San Diego County in 1868 to establish a cattle ranch, a venture in which Richard Deakers had made considerable investment, the couple made the decision to sell their Pittsburgh interests and move to California, where, as Nordhoff described it:

> There, and there only, on this planet, the traveler and resident may enjoy the delights of the tropics, without their penalties; a mild climate, not enervating, but healthful and health-restoring; a wonderfully and variously productive soil, without tropical malaria; the grandest scenery, with perfect security and comfort in traveling arrangements; strange customs, but neither lawlessness nor semi-barbarism.

Thus, in December of 1876, while her husband tied up matters in Pittsburgh, Catherine Deakers and her six surviving children, along with Uncle Hobson, left New York Harbor aboard the *Crescent City,* bound for California via Cape Horn, the Panama Canal being not yet in existence. Also traveling aboard the ship, though not listed on the manifest, were William and Hugh Patrick Mullholland, stowaways. Inevitably, the brothers were caught, and deposited on shore during a refueling stop at Colon, Panama.

Though William was confident that he and his brother could find employment aboard a California-bound ship leaving Balboa on the Pacific side of Panama, there was first the matter of making the forty-seven-mile crossing of the isthmus. Indeed a railroad could have provided transport, but the fare was $25, gold. Whether or not they had the money is unclear, but their decision was uncomplicated. They would walk the trail from Colon to Panama City. "That was the easiest way we knew to earn $25.00," Mulholland was fond of saying in his later years, "I would walk that far today to make $25.00."

Once the brothers reached Balboa, they quickly found employment on a Peruvian military ship, the *Bolivar,* sailing for Acapulco, where they worked for a month to help rig another ship, the *Frank Austin,* bound for San Francisco. "We landed in San Francisco in February 1877, sailing in through the Golden Gate on a spankin' breeze," Mulholland recalls. After a day or two's rest, they bought a pair of horses in Martinez and began the ride to Los Angeles by way of the San Joaquin Valley, much praised by Nordhoff for what he saw as unique drama: "The valley of the San Joaquin differs from an Illinois prairie in that it has two magnificent mountain ranges for its boundaries—the Sierra Nevada on the east, and the Coast Range on the west."

Something of what the brothers saw on their way to the Southland can be surmised by Nordhoff's description of the territory between Stockton and Merced: "Wheat, wheat, wheat, and nothing but wheat, is what you see on your journey, as far as the eye can reach over the plain in every direction. Fields of two, three, and four thousand acres make but small farms; here is a man who 'has in' 20,000 acres; here one with 40,000 acres, and another with some still more preposterous amount—all in wheat." Mulholland does not record what he thought of all that wheat, but he did say that the journey was inspiring. "I was tremendously interested in the whole country, although I can not recall definite impressions—everything was new, deeply interesting. The world was my oyster and I was just opening it."

As to his immediate impression of Los Angeles, he was even more fulsome. "Los Angeles was a place after my own heart," he recalled. "It was the most attractive town I had ever seen. The people were hospitable. There was plenty to do and a fair compensation offered for whatever you did." He went so far as to theorize that the place likely had the same attraction for him as it had for the Indians who had originally settled the area some 10,000 or 20,000 years before the arrival of Europeans.

Significantly enough, the Los Angeles River was the greatest attraction, Mulholland said. "It was a beautiful, limpid little stream with willows on its banks . . . so attractive to me that it at once became something about which my whole scheme of life was woven, I loved it so much."

For all its attractions, however, the still-rugged pueblo was in the grip of a smallpox epidemic, he and his brother quickly learned, and they were also dismayed to learn that two more of their cousins had died of typhoid while still aboard the *Crescent City*. Their young cousin Ella's own estimation of the place where her mother had dragged her was less than glowing: "Nothing looked like anything," she complained.

Furthermore, there was the matter of finding work, and Mulholland went a month without any. He was on his way to San Pedro Harbor to see about shipping out again, he told a *Los Angeles Times* reporter years later, when he chanced upon a man driving a well with a hand drill. The man—identified by Catherine Mulholland as Manuel Dominguez, grandson of the original grantee to Rancho San Pedro—offered Mulholland a job, and he accepted on the spot, thus transforming himself from a man who had made his living *on* the water to one who would define his life *with* it.

IN MYSTERY IS THE SOURCE

MULHOLLAND COULD SCARCELY HAVE KNOWN HOW significant his chance encounter with that well digger would be, and the fact is that he had one more adventure in him before he would cleave to the City of Angels once and for all. Beauty and mystical appeal aside, there were only 9,000 inhabitants in Los Angeles (San Francisco was then the country's tenth-largest city with just under 150,000) and, even with his gratitude for finding a job, the long-term prospects for a well digger's assistant could not have seemed utterly compelling for an ambitious young man.

Accordingly, late in 1877, William and brother Hugh Patrick set out on their final quest: a search for gold in the Arizona Territory. Ignoring reports that Apache tribesmen, including a band led by Geronimo, had undertaken a bloody series of uprisings in southern Arizona, the two made their way the 230 or so miles eastward from Los Angeles to Ehrenberg, just across the Colorado River from a pipe-dream settlement that would one day become Blythe, through which I-10 runs today. In Ehrenberg, the two got hold of a pair of burros and ventured into the barren mountains to the north, where a few placer claims had been filed.

According to the tale Mulholland later told his colleague J. B. Lippincott, the search for Arizona gold was fruitless, and the brothers turned their burros loose somewhere near the mouth of the Bill Williams River and floated down the Colorado the fifty miles or so back to Ehrenberg on a crude raft they had fashioned out of logs and reeds. At one point along the journey, the two ran out of food and pulled their raft to shore at the cabin of an enterprising settler who sold groceries to famished miners. When a fair number of halloos produced no response, the Mulholland boys entered the cabin to find it deserted. After some discussion, they took the groceries they needed and left, leaving payment on the table. It was the last they thought of the matter for years.

In 1924, nearly half a century later, William Mulholland traveled back to Parker, Arizona, another Colorado River town not far from where their unsuccessful prospecting adventure had taken place. Mulholland, who was in Parker as part of a survey for what would become another monumental water-diversion project, told the tale of his first venture to the area to a group in the hotel lobby where he was staying. When he had finished and was calling for another round of drinks, an old-timer seated amid the locals leaned forward.

"Well, I'll be damned," the old-timer said. "I've been wondering all these years who tracked up to the cabin from the river that day."

Mulholland stared back, puzzled.

"No way you could know this," the man said, "but the fellow who had that place was killed that day, right before you and your brother came in. Same time as you were in that cabin, his wife was run off into the hills with her baby, crying for help. Probably just as well you were gone when the posse showed up."

By this time, Mulholland, along with the rest of the audience in the room, was rapt. The old-timer waved his hand. "Of course they would have figured out it wasn't you. The old boy's brother was one of the posse, and he was a good tracker. They saw your tracks, of course, but they picked up the trail of the one who had done it and followed him all the way into Mexico."

If Mulholland was sitting there in a comfortable hotel lobby wondering if he'd miraculously escaped a lynching a half-century before, the old-timer dismissed it with a wave. "They brought him back for trial," the old man added.

It is only one of the numerous tales that swirl in the wake of the legendary Mulholland, so many of which seem to tread the border between myth and truth. Still, anyone tempted to doubt any Mulholland anecdote should keep in mind the indisputable facts of the man's legacy, which still keep a major city afloat, and which to this day provoke legislative debate, lawsuit, and acts of criminal mischief.

IN THE SPRING OF 1878, returned to Los Angeles from his prospecting trip, Mulholland took the job that, despite its unprepossessing aspect, would forever determine his fate—he became a ditch tender for the Los Angeles City Water Company, the privately held concern that had secured the lease to supply the city with its water. The water itself was diverted from the Los Angeles River at a point not far south of where today's Griffith Park butts close to Los Feliz Boulevard. That irrigation canal, dubbed the Zanja Madre, or "mother ditch," had been in use from the early days of Spanish settlement and, according to Spanish royal decree, the water in the canal as well as the river itself belonged to the pueblo that was situated upon it, a precedent that in the case of Los Angeles would eventually be upheld by the California Supreme Court.

However, the city had proved inept at the distribution of that water, and in 1868 the councilmen approved a thirty-year lease with the private consortium. In exchange for a promise to pipe water into domestic neighborhoods, install fire hydrants where necessary, provide free water to public buildings, and build an ornamental fountain in the city's plaza, the company would pay the city $1,500 a year, an amount that was later negotiated down to $400 a year. The city retained the rights to set annually the rates the company charged consumers for water, so long as those rates did not dip below those in effect in 1868.

Though the mayor opposed the plan and taxpayers were excluded from so much as voicing their concerns before the Council, the measure passed and while conniving and subterfuge would continue behind the scenes, as a public matter the privatization of the city's water supply was essentially settled for the remainder of the century. By the time Mulholland came on board, the company, in exchange for the rate reduction mentioned, grudgingly built the plaza fountain, constructed a new reservoir—the Buena Vista—to ensure a steady supply of water for the city, and ran a twenty-two-inch main all the way to the city center.

Still, much of the entire system's viability depended upon the Zanja Madre, and it was the newly hired Mulholland's job—at $1.50 per day—to keep it clean. According to another story passed along by J. B. Lippincott, Mulholland, shortly after being hired and twenty-three at the time, was down in the ditch, which ran alongside Riverside Drive, knee-deep in mud and shoveling for all he was worth to clear the flow. William Perry, then the water company's president, happened to be riding by on his way back to the city from the headwaters installation. Struck by the ditch tender's energy, Perry stopped to watch. Finally, Mulholland glanced up.

"Who are you?" Perry asked.

Mulholland scarcely paused. "It's none of your goddamned business," he replied and went back to his shoveling.

Perry simply rode on, but other workmen hurried over to tell Mulholland with whom he'd just been having a conversation.

"Is that right?" Mulholland said. He stuck his shovel into the mud, climbed the bank, and retrieved his coat, then made his way downtown in the wake of the departed Perry. He told a clerk in the company's office his name and that he was there "to get his time" before he was fired.

"You're the one from the ditch out by the river?" the clerk asked.

Mulholland nodded, not really surprised that word had already preceded him. The clerk rose and shook his hand.

"Mr. Perry says you're the foreman of the ditch gang out there from now on."

Mulholland stood by this story, which for him was profound. He'd behaved exactly as he was constituted, and instead of being punished, he'd been rewarded. To him it was an incident he often invoked to help explain a career.

The truth is likely more complex, of course, as borne out by a slightly different tale he told regarding his brief tenure as a well driller. That previous summer, before heading out to the Arizona territories, he was part of a crew hand-drilling an artesian well somewhere near Compton. They had gotten about 600 feet down, Mulholland recalled, when the bit struck the fossilized remains of a tree. A few feet farther down and other fossils turned up. "These things fired my curiosity," Mulholland said. "I wanted to know how they got there, so I got ahold of Joseph LeConte's book [likely the recently published *Elements of Geology*]. Right there I decided to become an engineer."

It is tempting to think of Mulholland's interest in the subject as drily encyclopedic, but a glance at LeConte's introduction suggests the true source of Mulholland's fascination. While the book would become an authoritative textbook, LeConte, cofounder of the Sierra Club with John Muir in 1891, insisted on greater aspiration for his work: "I have been guided by long experience, as to what it is possible to make interesting to a class of young men, somewhat advanced in general culture and eager for knowledge, but not expecting to become special geologists. In a word, I have tried to give such knowledge as every thoroughly cultured man ought to have."

While that might seem an ambitious stretch for modern readers, it is easy to imagine that Mulholland took LeConte at his word when the good professor insisted that his was a dynamic subject. "Geology may be defined, therefore, as the history of the earth and its inhabitants, as revealed in its structure, and as interpreted by forces still in operation."

Judging by Mulholland's later actions, the interest in the interplay between the physical world and its inhabitants could in fact be the key to understanding what would eventually distinguish Mulholland from many of the highly trained engineers with whom he would labor. As to whether or not the acquisition of LeConte's book was the beginning of Mulholland's training in the field, it is undeniably accurate in characterizing the nature of that training. For all that he accomplished, and though he would one day be inducted into the American Society of Civil Engineers and be granted an honorary LLD from Cal-Berkeley, all of Mulholland's engineering training came either on the job or through his own reading.

If his restlessness had been an impediment to him as a schoolboy, Mulholland carried an understanding of the importance of reading throughout his life, often telling associates that he could not fathom a man who did not read. "I have always had a good memory as well as a nose for news, everything impressing me in some way and hooked up in my mind with important events and dates of history," he would write in his sketch.

Thomas Brooks, a coworker of Mulholland's from their earliest days with the water company, recalled that during a period when they were rooming together, Mulholland would stay up long after Brooks had fallen asleep, poring through works as diverse as John Thomas Fanning's *Practical Treatise on Hydraulics and Water Supply*, John Trautwine's *Civil Engineer's Pocket Book*, and Shakespeare. In addition, Brooks said, Mulholland was also "fond of Grand Opera."

Once, when he was being cross-examined as an engineering expert, an attorney hounded him: "Will you please tell us what education qualifies you on these matters?"

Mulholland did not hesitate. "I learned the three R's and the Ten Commandments, received my mother's blessing, and here I am."

Among the benefits of Mulholland's new status as ditch-gang foreman were quarters provided by the company, or what he called "a shack near the Old Sycamore Tree" on company land near the diversion point of the Zanja Madre and the then tiny Buena Vista Reservoir; his love of the place and the work are evident in his own descriptions. He would rebuild and expand the reservoir—the principal water storage source for the city until 1902—in the years to come, planting more than 1,000 trees, "with my own hands . . . still growing there, blue and flaming eucalyptus, desert palms, and oaks."

The Zanja Madre diverted water from the Los Angeles River at the North Broadway Bridge into four main lines, branches that Mulholland said "really beautified the town." Though the mother ditch was a formidable canal, Mulholland said of the branch lines that "a sturdy man could jump across one." In those days, he recalled, "The women used to wash clothes at certain places in those zanjas, wading in and clouting the clothes on rocks. I was fighting this use of the zanjas until 1902."

He lived in his shack in the Elysian Hills for two years, laboring, reading, and always inquiring of the trained surveyors and engineers with whom he worked as to best methods and practices in hydrology. Given his down-to-earth nature, local ranchers and developers grew to trust him.

In 1880, he was promoted from ditch-tending duties and was placed in charge of a pipe-laying crew, his pay rising to $65 a month, though he had to furnish his own horse. While he enjoyed the raise and the move to a higher grade of shack that came with the promotion, Mulholland was also buoyed by the heightened technical aspects of the new work and the increased contact with principals in the company, particularly superintendent Fred Eaton, with whom he would develop a storied association.

In 1886, with Fred Eaton having resigned from the water company to become city engineer for Los Angeles, Mulholland was assigned the oversight of what was the most ambitious construction project his company had overtaken, the building of a five-mile-long conduit to enclose the Los Angeles River as it descended through the narrows below Griffith Park. Designed to protect the water from the polluting effects of the ever-encroaching settlement growing about it, the conduit was built of wood and lined with concrete and proceeded largely without incident under Mulholland's direction.

Then, in November, the superintendent who had been hired to replace Fred Eaton died of a heart attack. Water company president William Perry offered the job to Thomas Brooks, who had risen to the post of assistant superintendent. But Brooks, only twenty-four and sensing that the position was beyond his capabilities, declined. Why not give the job to Bill Mulholland, Brooks suggested? There was a man who knew water.

Mulholland accepted, as we know, taking on at thirty-one a position he would hold for most of the rest of his life. In the early years, he saw the city's population grow to nearly 50,000 and endured complaints of fishy-tasting water and pressure sometimes too paltry to feed fire hoses. In 1888, the company built a new headquarters where he and Brooks were roommates for a time, and in 1890 Mulholland got his first brush with celebrity when company president Perry identified him to the *Los Angeles Times* as the man who had averted a total shutdown of its water supply when a flash flood clogged the main conduit at the Crystal Springs headwaters. Mulholland had worked without sleep for four days, Perry told reporters, unclogging the conduit and restoring the flow of the water before any citizen was aware of the incident. For his part, Mulholland said, "I never had my shoes off from Tuesday until Friday night."

Mulholland got a gold watch from the company in return for his stalwart services, but at about the same time, he received a much more significant prize when, despite his own protestations that while he "enjoyed a wide interest in all phases of experience except girls—I was never known as a lady's man or for feminine accomplishments," he met and won the heart of Lillie Ferguson. Mulholland met Lillie, then twenty-one, in the summer of 1889, when he knocked on the door of her father's home while seeking permission for a surveyor's encroachment onto the Fergusons' Los Feliz property. She would later write that from the moment her eyes met with blue-eyed Bill's, she knew he was the man she would marry, and on July 3, 1890, she did just that.

Not much has been told about Lillie, who did little socializing during their twenty-five-year marriage, perhaps because she had suffered mild hearing loss as a result of a childhood illness. She grew up in Michigan, not far from the lumber camps where Mulholland had nearly lost a leg, attended convent schools, and came west with her parents in 1887 not long before she met Mulholland. Though she was said to be quiet and retiring, Lillie by all accounts lived a fulfilling life with her children as her jewels. Lillie's shyness might also have been explained by the dynamic character of Mulholland's new mother-in-law, who was anything but a wallflower. Though she had been christened Frances (she was the ninth of ten children), she preferred to be called Frank and in fact had her silver wedding spoons engraved "Frank Ferguson." She wore a nightcap, smoked a pipe in bed before she went to sleep, and claimed psychic powers, often stopping the preparation of meals (said to be prodigious) with the proclamation that some unplanned-for guest was about to arrive.

Following the 1890 ceremony, the couple moved into a house at 914 Buena Vista, the first real home Mulholland had experienced since his arrival in America more than sixteen years before. From logger to well-digger to prospector to ditch tender to superintendent of a city's water company, William Mulholland, with a wife and a proper home at last, had experienced a great deal in a relatively short time, and he was still shy of his thirty-fifth birthday. And he would, in short order, experience a great deal more as his first three children were born: Rose in 1891, Perry in 1892, and Thomas in 1894.

"Perry," as he was generally known, was actually christened "William Perry" as a result of a bit of typical Mulholland blarney. In a tale that is related in her manuscript, Catherine Mulholland explains that when her grandfather showed up at work to announce the good news of his son's birth, company president William Perry asked Mulholland what he had named the boy. "Well, we're going to name him after you, Bill," Mulholland joked. When a set of silver flatware engraved "WPM" arrived at the household on the following day, Lillie, who intended to name her son after her father, James Ferguson, asked Mulholland what was going on. Mulholland was forced to tell her what he had blurted out to his boss and convinced his wife that none of it could be undone. Lillie went along, but for years referred privately—and perhaps possessively—to her son as "Boy Blue."

Yet for all that had taken place, Mulholland could have scarcely imagined what would unfold in the second half of his life, including a series of events that would eventually pit him against his employers and change virtually everything about the nature of his work before the ensuing decade was out. While he once wrote, "From the time I became superintendent and general manager of the water company in 1886 until the latter years of the century there is nothing special to report," it seems impossible for Mulholland to have missed the portents of what was about to come.

WHOSE WATER IS IT ANYWAY?

BY 1890, THE FRONTIER TOWN OF 9,000 THAT MULHOLland had encountered in 1877 was greatly transformed, its population grown to nearly 50,000. Law and order had come to a city where at the time of his arrival, disputes were still being settled by gunfights in the streets and the social order maintained by lynchings. A county bar association was established in 1878 and the University of Southern California was founded in 1880.

The Southern Pacific Railroad extended its line from San Francisco in 1876, linking Los Angeles to the rest of the nation for the first time, and by 1885, the Atchison, Topeka, and Santa Fe had come to town as well. In 1887 when the Santa Fe began lowering the price of a ticket from Kansas City, Missouri, to Los Angeles, first from $125 to $15, and then, on March 6, to $1, it fueled an enormous jump in tourism and relocation. The days of the wagon trains were long gone, and it was no longer necessary to brave a months-long voyage around Cape Horn or even to find significant wherewithal for the train ticket necessary to try one's luck in a land where oranges grew year round. For a dollar, you could hop on the train to paradise.

From the time of Mulholland's arrival, agriculture in the outlying areas of the city had steadily encroached on ranching, with neatly tended groves and orchards displacing pasturelands. Owing to the development of refrigerated boxcars, the same rail lines that brought new settlers in were also carrying the newly hybridized navel and Valencia oranges back east, along with other crops. Present-day Beverly Hills was thriving in the 1880s, principally as a vast bean field. The area just south of Cahuenga Pass that would become Hollywood was a fig orchard.

The newspaper that would become dominant in the city had begun operations as the *Los Angeles Daily Times* in 1881 (Harrison Gray Otis would take over as editor and publisher in 1882). Along with stories that the city's Conservatory of Music (one day to be folded into Disney's CalArts) was founded in 1883 and that in 1889 the USC "Methodists" had drubbed St. Vincent's College (later Loyola Marymount) 40–0 in the first collegiate football game played in Los Angeles, the paper—in competition with such rivals as the *Herald*, the weekly *Porcupine*, and the short-lived *Tribune*—was running advertisements in 1890 that Prince Albert men's suits could be purchased for as little as $12.50, Johnston & Murphy shoes for $6, and derbies for 95 cents. Prime orange-growing land was available for $125 to $200 an acre, and Mulholland could have purchased a home site in the Los Feliz Ranchos from his developer father-in-law for anywhere from $150 (100-foot frontage) to $750 (five acres).

While the startling changes in early Los Angeles are diverting to look back upon, for William Mulholland such development translated principally into difficulties at work. More people and more businesses simply meant more problems for the water company. There were periodic complaints from citizens about cloudiness or a "fishy" taste in their tap water and reports in the papers that by the end of the day, upper floors in downtown buildings (at the end of the water mains) often had no water pressure at all. There were other reports that firefighters in outlying districts often found only a dribble issuing from hydrants and were helpless to save homes and businesses.

Most of the issues had to do with the historically uncertain rate of flow from the Los Angeles River. The river itself begins officially in the west San Fernando Valley near Canoga Park, where Bell Creek, flowing eastward from the Simi Hills, joins with the Arroyo Calabasas, which flows northward, draining a portion of the Santa Monica Mountains. From there the river flows for some forty-five miles eastward through the San Fernando Valley until it joins with the Tujunga Wash near present-day Studio City. Tujunga Wash drains about 225 square miles of territory in the San Gabriel Mountains east of the San Fernando Valley and carries significant water only in the rainy winter and early spring months. From this juncture, the river bends southward around today's Griffith Park, passes through the Glendale Narrows, and continues on for a bit more than thirty miles to its mouth at Long Beach.

From the earliest years of settlement, the Los Angeles River was the primary source of water for inhabitants, who, whether Indian, Spanish, or greenhorn settler, had to put up with seasonal scarcities interrupted by times of torrential flooding. From his earliest days with the water company, Mulholland was well aware of the city's dependence on this unreliable source, but there seemed little he could do about the matter other than improving storage reservoirs, avoiding waste, and preaching conservation. However, with the population having quintupled since his own arrival and with no end in sight, it was becoming clear that such tactics had their limit—something would have to change or Los Angeles would simply cease to grow.

Mulholland was not the only person to realize that some sort of impasse was imminent. As the days began to count down to the expiration of the water company's lease in 1898, some civic leaders began to campaign for the city's reappropriation of its most valuable resource. As early as the end of 1891, papers carried news that the city had approached William Perry to ask at what price he might be willing to sell. Perry's quick tally of assets resulted in an estimate of $2.5 million, though he cautioned that he was not making any offer. Such a matter would have to be taken up by his board in consultation with their stockholders. Nonetheless, his figures were dismaying to politicians who could only imagine the level of resistance they would encounter among taxpayers who would have to foot such an astronomical bill.

Some critics countered that the city should simply build its own waterworks and bypass the company altogether, but the courts advised that under the terms of the lease, only the Los Angeles City Water Company was authorized to sell the river's water to the citizenry. Furthermore, the founders of the company had anticipated that the question of ancient pueblo rights to the water would likely someday come under challenge. In the British legal system, which guides the principal practice of law in the United States, the concept of riparian rights prevails—basically, whoever owns the land through which a river flows owns the water flowing therein. Under that concept, the city was entitled only to the water from the part of the river that flowed within the city limits.

With this in mind, the Los Angeles City Water Company, shortly after signing the original lease with the city, had created a separate corporate entity that they called the Crystal Springs Land and Water Company. The chief asset of this subsidiary consisted of a sizable tract of marshland adjacent to the Los Angeles River just above the Glendale Narrows from which a groundwater supply would be extracted, water that ostensibly was not part of the Los Angeles River. As a further attempt to distance its operations from city influence, the water company then transferred its own interest in the city's river water to its own subsidiary for the standard $1 and other considerations. Eventually, the Crystal Springs Land and Water Company would contend that, as a separate corporate entity, it was not bound by the terms of the original lease; but, given that the two concerns shared a common board of directors and set of officers, this argument did not prevail.

Still, if the city wished to purchase the Los Angeles Water Company, they would have to purchase the holdings of Crystal Springs as well, and by 1893 the price had risen to $3.3 million. The city had meanwhile commissioned its own estimate of the value of the water company holdings, which was dutifully reported in the newspapers. According to City Engineer Henry Dockweiler, the appropriate sum was just shy of $1.5 million. At that point, the matter became moot when the company responded that Dockweiler's figures did not take into account the value of the water sources the company had developed at Crystal Springs and that, furthermore, the company was not for sale at any price.

Mulholland, as superintendent of operations for the company, was not involved in questions of corporate policy. His job was to keep the pipes flowing as best he could, a position that suited him perfectly, for he was a doer and not a paper-pusher, his attitude toward such detail best summed up by the observation, "If you leave a letter in the basket long enough, it will take care of itself."

Yet he was increasingly viewed as the face of the Los Angeles

City Water Company and late in 1892 had been called upon to testify on behalf of the city and the company in what would turn out to be a significant court case, *Vernon Irrigation v. the City of Los Angeles,* involving the city's rights to river water. In that case, a pair of developers argued that they were entitled to withdraw a certain portion of water from the Los Angeles River to benefit their new development just south of the city's center. The judge in the case eventually rejected the plantiffs' claims, affirming that the city owned the water in the river and though it was entitled to sell any excess water to any parties it wished, it could not be compelled to do so.

That might have ended the case, but the plaintiffs appealed to the California Supreme Court, which ruled in 1895, affirming the original decision in the main. However, the Supreme Court took exception to the notion that the city was free to sell any excess water for the purpose of irrigating land outside the municipal boundaries. At the time, no one gave much thought to this ruling, for the city was quickly approaching the point where the notion of "excess water" was a fantasy. As history demonstrates, however, such afterthoughts have a way of reinserting themselves into the greater scheme of things.

In some ways, Mulholland's insistence that there was "little to report" regarding his first decade as superintendent of the water company was truthful. Though he would have had his own opinions about the men for whom he worked, he was above all a loyal man. He had his share of confrontations with Perry and the board of directors, but they were primarily concerned with practical issues of maintenance of the waterworks that often conflicted with Perry's desire to increase profits: the costs involved with improvement of mains, expansion of storage reservoirs, and maintenance of fire hydrants and street-sprinkling valves.

If there was one bedrock conviction that Mulholland held and that conflicted with those of his superiors, it was that their customers were owed what they were paying for. Nor was he trying to squeeze blood from a rock; the company's revenues had grown from $20,000 in its first year to $425,000 in 1896. On the other hand, he viewed the notion that the city could duplicate what had become a vast and basically sound water-delivery system with a parallel system of its own as lunatic at best and political grandstanding at worst. From a system that began with 3 miles of wooden pipe and 1 mile composed of iron, by 1896 there were 325 miles of water pipeline laid within the city. To Mulholland, what would happen in 1898 remained to be seen. Until then, a deal was a deal.

Meanwhile, pressure from the city on Perry and the company's board increased. In 1897 Mayor Meredith Snyder addressed the City Council, lamenting that there had been much talk but little action in moving matters forward toward some resolution with the water company. "Each citizen demands that something be done that he be protected from further suffering caused by a corporation whose greed is great and relentless," Snyder said. The company was holding out for $3 million, and the city's offer was firm at $1.3 million.

Finally, in July 1898 a board of arbitration was appointed, composed of one representative chosen by the city, a second put forward by the water company, and a third elected by the other two. The city also appointed a board of engineers deemed better suited to supporting their valuation of the company's holdings to the arbiters. As J. B. Lippincott, a member of the city's engineering board recalls, they of course called in Mulholland, as chief engineer and plant manager, for questioning.

Mulholland was his typical no-nonsense self when he took his seat before the board. What exactly was it that they wanted to know? he asked, in a tone that suggested he could be somewhere getting real work done.

The board needed a complete list, engineer James Schuyler replied. "The length of pipe, its size, character, and age. We also want to know the number of gate valves and all about them, as well as the number and position of fire hydrants and all other structures connected with the water system."

Schuyler and the others presumed that their request would precipitate a delay of indeterminate length while the superintendent busied his staff with the collection of the necessary documents, essentially stalling the entire process. Still, they were determined to have a firm basis upon which to press their case with the arbiters.

Thus Mulholland's answer was something of a stunner. "Get a map," Mulholland told the board, "and I'll show you."

The board had a drafting table cleared and a sizable plat of the city spread out. Mulholland then proceeded to sketch out from memory a diagram of every pipe on every city street, along with notations of size, type, and age, as well as a description of connective fittings, the location of gate valves, and fire hydrants. The board sat in amazement as Mulholland worked, but still, Schuyler pointed out at the end of the process, "memory" did not seem an adequate basis upon which to proceed in such a matter.

Mulholland stared back. What then, would the board have him do? After a conference, the board reached a consensus. If Mulholland was not opposed, they wished to dig up a few places along the lines that he had sketched out on the map, just to check.

Mulholland did not mind, he assured the board with a wave. Where exactly would they like to dig?

If Schuyler was smiling smugly as he made notations on the map, the record does not reflect it. In the end, the board placed red circles on 200 locations scattered about Mulholland's map. One could assume that Mulholland would be affronted or at least exasperated by such a request, but as Lippincott recalls, the superintendent actually seemed cheerful as he left the meeting.

Two hundred excavations were subsequently made, Lippincott reports, and at each the findings bore out Mulholland's notations to the letter. It is a story often repeated, but one that reflected typical Mulholland practice, says Lippincott, to allow interrogators and opponents to exhaust themselves while he placidly endured, waiting for the right moment for one solid counterpunch. Nor was his unwillingness to suffer fools limited only to his adversaries. During the protracted arbitration proceedings that followed, Mulholland sat at his employer's table while a water company attorney badgered a city witness unmercifully.

Finally, the attorney, losing steam but still determined to discredit the witness, turned to Mulholland for help. "What else shall I ask him?" the attorney asked quietly.

Mulholland leaned close. "Ask where he got that red necktie he is wearing," the superintendent whispered.

The attorney glanced hurriedly at the witness, then back at Mulholland, who was now sitting sphinxlike. Quickly the attorney rose and asked the judge for a recess, which was granted.

"What *about* that necktie?" the puzzled attorney asked Mulholland when they were alone.

The water superintendent shrugged. The attorney had already asked everything else in the world. It was the only question he could think of.

A CIVIL SERVANT BORN

DESPITE THE FACT THAT THE COMPANY'S LEASE EXpired on July 21, 1898, it would take nearly four years for negotiations between the Los Angeles City Water Company and the city to play out. In the meantime, the city's population more than doubled between 1890 and 1900, to 102,479, and Fred Eaton, former superintendent of the water company and Mulholland's boss, had been elected the city's mayor as a business-friendly candidate dedicated to the development of Los Angeles. Eaton was no foe of private enterprise, to be sure, but, as the descendant of a land-owning family and well versed in the importance of water to survival in the West, he knew that the city could not be held hostage to private interests when it came to water. From the time of his election in 1898, he devoted a large part of his efforts to securing the takeover of the water company—some in his family say that it was the only reason he ran for office.

Despite Mulholland's meticulous inventory of the company's

holdings, and over the protests of the company's representative to the arbitration panel, the city determined the value of the waterworks as slightly more than $1 million. The company countered that the works were worth $2 million and that their holdings at Crystal Springs were worth another $1 million. The impasse began to crumble in 1899, when the California Supreme Court ruled in a separate suit (*City of Los Angeles v. Pomeroy*) that the underground waters of the Los Angeles River running above the Glendale Narrows were indeed part of the city's property and that developers had no right to siphon off those waters in any fashion.

Following the court's decision, Fred Eaton called for a special election to approve $2 million for the acquisition of the water company's holdings and improvements to the system. The bond issue passed overwhelmingly, but the company in turn filed suit, arguing against the amount of the arbitration panel's valuation of the waterworks, and as the fight continued, the impracticality of selling the bonds during pending litigation scotched the process.

Meantime, Eaton and J. B. Lippincott engaged in attempts to prove, once and for all, that any waters draining into the San Fernando Valley from its surrounding watershed, whether flowing on the surface or into underground chambers, were essentially part and parcel of the Los Angeles River. Eaton had, with the assistance of Thomas Means, a Department of Agriculture soil expert, "established the sources of the various underground feeders of the Los Angeles River," the *Los Angeles Times* reported. An analysis of water samples taken far up the San Fernando Valley showed them to be identical with those taken near the Glendale Narrows. The mayor told reporters that he "therefore knows the exact source of the water supply of the city."

Other court proceedings established that, ownership of flowing water underground aside, the "lease" that the city had signed with the water company in 1868 was still in effect and was a binding contract that obligated the city to pay the company a fair price for its holdings. To complicate the situation further, all of this wrangling was taking place against a backdrop of a drought that had persisted since 1892, leading to the company's request that water rates be raised for the first year of the new century, and underscoring the public's anxiety that its interests were being ignored.

Finally, in July 1901, with Eaton out of office and the flow in the Los Angeles River nearing a three-year nadir, the city made another offer to the water company: $2 million for everything, including the holdings at Crystal Springs. Banker I. W. Hellman, founder of the company that would become Wells Fargo and chief stockholder of the water company at the time, called William Mulholland to his summer home in Lake Tahoe to ask what his superintendent thought of the deal.

Mulholland would later explain that he had found the various contentions of politicians regarding the true value of the waterworks to be principally self-serving appeals to the electorate. In any such business transaction, he said, taxpayers would have to expect to pay "for at least some bone with the meat." In a speech to the Sunset Club, he gave his opinion that the compromise was a wise one, putting an end to a situation that jeopardized the city's progress.

It is also likely that the pragmatic Mulholland cautioned Hellman that it was possible that the company's various court challenges and associated legal tactics would someday come to an end and that there might be a condemnation ordered that would yield the company little. Whatever he told Hellman at Lake Tahoe, the outcome is clear. On July 29, Hellman informed the city that the company would accept the $2 million offer.

An election for a new bond issue was quickly arranged, and despite some carping that the payment would result in a $1 million "steal" by the owners of the company, the matter was decided by a five-to-one margin. By the end of August 1901, it seemed certain that the city would finally have its water to itself.

Though Eastern financiers were generally optimistic about the prospects of development in Los Angeles, an economic downturn slowed the sale of the bonds, and the threat of a challenge to the election had been rumored. Still, city attorney William B. Mathews was persuasive in convincing Eastern banking interests that the issue was a gilt-edged opportunity. Though Mathews and city treasurer William Workman would have to spend most of the winter at the task, by late February the bonds were sold, the monies transferred, and the water company became officially the property of the City of Los Angeles. Workman's return, covered by the *Los Angeles Times* on February 11, was marked with a parade led by a brass band, so great was the relief of city boosters. "Wherever I went the mention of California, and especially of Los Angeles, attracted attention," Workman told reporters before giving full credit for the successful sales pitch to Mathews, without whom nothing would have happened. It was a theme that Mulholland would often repeat in the years to come. He had indeed done the work, Mulholland liked to joke, but it was Mathews "who kept me out of jail."

There was no controversy concerning the proposed governance of the new agency. The City Council was in agreement that an elected board of water commissioners would oversee the system, which would operate under civil service rules. All thirty-one employees of the private company would be retained. Only one question remained: Who would lead the new city department?

It was not much of a question in reality, although there was one minor sticking point. As superintendent of the private company, William Mulholland's annual salary had risen to $3,000. When he appeared before the City Council to discuss the new position,

Mulholland explained that he would need a raise to $5,000. When some council members quibbled, Mulholland explained that as a private employee, he had been free to take on virtually any consulting job he was offered. In 1900, he said, he had made $8,000 doing such outside work. The city required that he devote all his time to the new position. In the same spirit of compromise that had dictated the purchase of the private company, the City Council agreed to Mulholland's terms. As Mulholland later liked to joke, "When the city bought the works, they bought me along with it." At forty-seven, the Chief became a civil servant.

In his first appearance before the temporarily appointed board of water commissioners, Mulholland stressed the need for repair and extension of a system that he said was already a decade out of date. There were two principal items on his agenda: the construction of a vast underground storage facility above the Glendale Narrows that would capture more than 6.5 million gallons of hidden water per day, and the somewhat startling intention to install meters on every customer's line. As he told the *Times* on May 23, shortly after his appointment, taxpayers should not worry—his research had showed him that in communities where water meters were being used, water bills had often been cut in half.

While modern readers possibly view water meters as being as inevitable as death and taxes, the devices were still somewhat newfangled at the turn of the twentieth century. The concept of metering water by volume was kicked about in Roman times, and Leonardo da Vinci designed a prototype in the sixteenth century. But it was 1855 before the first US patent for a reliable device was issued to a man named Henry Worthington, and, according to a staffer at the Smithsonian, one hydraulic engineer was complaining in the pages of *Scientific American* as late as 1870 that "Measurement of water flowing through pipes, under any and all circumstances of position, pressure, and velocity, has, perhaps, more difficulties than any other with which the modern mechanic can grapple."

Mulholland was resolute regarding the need for a metering system, however, in large part because he was convinced that the devices would in the end reduce consumption, which in 1902 stood at 267 gallons per person daily (the city's first water meter had actually been installed at Stern's Winery by Thomas Brooks in 1889). Based on the rate of current population increase at that time, Mulholland estimated that demand would rise to 27 million gallons daily, while only 23.5 million were available from present sources. He planned to make up for some of the shortfall from the new subterranean reservoir already under construction, and he could always pray for rain, but finding a way to reduce the city's thirst, and certainly its waste of water, was paramount.

Mulholland was well aware of the natural limitations against which he labored. There was no dependable snowpack feeding the 834-square-mile watershed of the Los Angeles River, and the long-term drought in the region showed no signs of abating. There had been steady decreases in the flow every year since 1893, he wrote in an early semiannual report, and in 1902 the river was carrying less than half of the volume it had in 1893.

W. C. Mendenhall, an official of the US Geological Survey, had opined in the late 1890s that the waters stored in the natural reservoir beneath the San Fernando Valley were sufficient to supply the city with water for seven years if not a single drop of rain was to fall in all that time. However, the population had doubled since the time those brave words were uttered. With more than 100,000 in Los Angeles and no end of settlers in sight, a huge shadow had drifted over the prospects for the boundless development foreseen by Fred Eaton and others.

Still, Mulholland spent the first four years of his tenure with the newly christened water department doing what he could. Following up on his conviction regarding water meters, the department reduced rates by 50 percent to every customer who agreed to such an installation. As a nod to the notion that the water company stockholders had been riding the coattails of taxpayers for too long, rates were reduced 10 percent across the board, with no evident ill effects upon the department. After paying all operating expenses, including the interest and principal on the bonds for the purchase, the department had realized a net profit of $1.5 million, which, much to Mulholland's satisfaction, was plowed back into operations instead of stockholder profits. In addition, the installation of the meters had also reduced per capita daily consumption, which had previously risen to among the highest in the nation at 300 gallons daily, by one third.

Accolades flowed Mulholland's way, with one writer declaring, "Superintendent William Mulholland, who had spent most of his mature manhood in the employ of the Water Company, has been a veritable tower of strength, giving to his duty without doubt far more freely from his energy of mind and body than he could possibly have done if he had owned the whole plant with the profits flowing into his own pockets." But nothing seemed sufficient to meet the ever-growing demand. Mulholland had in his early days with the department estimated that at maximum efficiency and minimum waste, all the water resources of the existing watershed would be able to support a population of 250,000. When he made that prediction, he assumed that it could be well into the second decade of the century before a final solution would have to be addressed. But as 1905 approached, the city had nearly doubled in size again.

Meantime, the city charter was amended to provide that the five-member board of water commissioners be appointed by the mayor instead of being elected, a step that was seen as a protection against politics intruding in issues impacting the city's water supply. However, there were unintended consequences, for the amendment gave the commissioners virtually free reign over the water department and its finances. At the same time, another amendment to the city charter was enacted that prohibited the sale or transfer of any right to water "now or hereafter owned or controlled" by the city without a two-thirds vote of the citizenry. While the intent of the amendment was to ensure that the city's water supply would never again fall into private hands, it too would prove far more significant than the city fathers could have foreseen.

All the while, speculators continued to invest in outlying lands and planned developments, thereby ensuring that the problems facing Mulholland and his new department would only escalate. In April 1903, Theodore Roosevelt, who had assumed the presidency when William McKinley was assassinated in 1901, came to the city to proclaim it "a veritable garden of the earth." In September of that same year, a syndicate that included General Harrison Gray Otis of the *Los Angeles Times,* as well as streetcar magnate Henry Huntington, Union Pacific mogul E. H. Harriman, and other influential Los Angeles businessmen, announced the purchase of the 16,540-acre ranch of George K. Porter, a tract that constituted nearly 10 percent of the lands in the San Fernando Valley. The acquisition was described to the *Times* as being made for the purposes of developing the property, though without access to water, substantial development seemed unlikely in that far-flung, north-central section of the valley.

Meantime, through the remainder of 1903 and well into the normal rainy months of 1904, the drought persisted. The Geological Survey's Mendenhall reported that underground resources were drying up at catastrophic levels. Lands where wells could be successfully dug in the San Fernando Valley had decreased by one-third, and one major source—the Bouton Well—which in 1899 had produced 4 million gallons a day, had dropped to less than 800,000 gallons. The water level in a major well near Anaheim, located above the largest underground reservoir in Southern California, had plunged from 23 feet below the surface in 1898 to 52 feet in 1904.

Cattle were starving on ranches in the Antelope Valley, Lake Elizabeth had dried into a mudflat, and by July the city was consuming more water than was flowing into the storage reservoirs. Mulholland declared a ban on watering lawns and turned off the flow to the ponds in the city's parks, all the while spurring his men to bring the pumps at the new underground reservoir above the Glendale Narrows on line. Once those pumps were feeding the system and providing additional waters to be stored at the Buena Vista Dam, Mulholland believed the city, with its thirst risen to 33 million gallons per day, would make it through the crisis.

Finally, in July, the new pumps were brought on line, but Mulholland quickly noticed something odd. Within two weeks, and despite all ongoing conservation efforts, the city's consumption had risen to more than 40 million gallons per day. Even at night, when consumption was normally at its lowest, the reservoirs were failing to refill. Somewhere, something had gone very wrong.

Though it took a week's investigation, the answer finally became clear. The water being sucked up from the new underground gallery and packed into the aging reservoirs as excess was being discharged through faulty gates and valves into the system's overflow sewer lines, Mulholland told reporters. Nine million gallons of water each day were lost through the Cudahy Sewer (draining the Boyle Heights district) and others in the aging sewer system. (Using storm drains to dump water into the ocean would become a central plot device in *Chinatown,* where the "loss" was portrayed as purposeful, part of a plan to panic citizens and drive support for the building of a new dam.) Though he was able to put a stop to the leaks and a freak summer rainstorm brought some relief to the city, Mulholland was nearing the end of his rope.

Though he had turned down the commissioners' suggestion that he take a vacation earlier in the year, in September he asked for a leave of three weeks, a request that the commissioners, grateful for his stolid service, were happy to grant. If any member of the board was suspicious of the true purpose of the Chief's request, no mention was made.

ROAD TRIP

MULHOLLAND HAD ALREADY CONFIDED TO ASSOCIates that the Los Angeles River, whose resources he had been so confident of just three years before, would essentially be tapped out, and soon, at a flow of 46 million gallons per day. Per capita water consumption had been reduced to 144 gallons daily, less than half of what it had been when he took over, but he despaired of being able to improve much upon that figure. Further, he estimated that the rise in population had far outstripped his estimates. In four years, there were already more than twice as many residents as there had been in 1900. There would be a quarter million by 1905, nearly half a million by 1915, and 700,000 residents by 1925.

As he liked to joke, they could either kill Frank Wiggins, principal spokesman for the development-happy Los Angeles Chamber of Commerce, or face the facts: the river was outmatched and the city would either have to accept a cap on population, or it would be forced "to supplement its flow from some other source." The burning question was what that source would be.

With the assistance of engineers J. B. Lippincott and O. K. Parker, Mulholland had completed an exhaustive survey of all existing water resources in Southern California. It was his conclusion, issued as a part of his "Fourth Annual Report of the Water Commissioners," that any attempt to develop storage and supply facilities on any stream that flowed anywhere south of the Tehachipi Mountains, that is, the northernmost boundary of distant Antelope Valley, would in essence rob water from the watershed of the San Fernando Valley. Some advocated building a reservoir to capture floodwaters of the San Gabriel River in the southern part of the county near Azusa, but Mulholland doubted that there was a viable site for a reservoir there, and Mendenhall's calculations indicated that this intermittent source would fall far short of what was needed.

Others suggested tapping the Mojave River, another intermittent stream on the northern side of the San Bernardino Mountains, but again Mulholland cited an insufficient rate of flow and an inordinately high cost of impounding and pumping these waters to the city. Either such source would provide only about one-tenth of what would be needed to support the city's growth to 1 million, a figure that he believed to be a farsighted benchmark for the future. But there simply was no such source of water available in Southern California.

The city was indeed in a quandary, one that finally led Mulholland to send a long-delayed message to Fred Eaton. The Chief was following up on a matter that Eaton had been floating for years; it had been conveyed to the former mayor in September 1904. Perhaps it was an idea whose time had come.

The story of exactly where and under what circumstances Eaton and Mulholland met to discuss the Owens River possibilities in 1904 has been told and retold many times, including various versions from the principals themselves. But all agree on the key points, and Mulholland was ever generous in crediting Eaton with the germination of the idea. "Thirteen years ago Fred Eaton first told me that Los Angeles would one day secure its water supply from the Owens Valley," Mulholland told a reporter for the *Los Angeles Times* in an interview shortly after the project was announced. "At that time the Los Angeles River was running 40 million gallons of water daily and we had a population of less than 50,000. I laughed at him." By Mulholland's reckoning at the time, the Los Angeles River was capable of supplying the city for a half-century to come.

But Eaton told Mulholland he was being shortsighted. "You have not lived in this country as long as I have," he said. "I was born here and have seen dry years, years that you know nothing about. Wait and see."

Well, Mulholland told the reporter, he had waited and he *had* seen. "Four years ago I began to discover that Fred was right. Our population climbed to the top and the bottom appeared to drop out of the [Los Angeles] river."

When the two met in September 1904, Eaton reiterated his long-held certainty that the solution to the water problems of Los Angeles was to be found in the Owens Valley; shortly after that meeting, Eaton and Mulholland hired a mule team and a buckboard, and with that legendary cache of whiskey packed among their supplies, they set out on a 250-mile trek that would change history.

Aside from the discarded whiskey bottles that purportedly marked their trail, the pair did not closely document the original trip from Los Angeles to the Owens Valley that they took in September 1904. However, Eaton did take notes on a follow-up fact-finding mission about a year later, one that would have followed an identical route. Eaton also took a number of pictures of the rugged trail and settlements along the way and included them in a commemorative album of which he gave Mulholland a copy that is still in Department of Water and Power files.

The second time around, Eaton had in tow reporters from the *Times,* the *Express,* and William Randolph Hearst's new *Examiner,* founded to help advance Hearst's presidential ambitions in 1903. Also along for the trip were the city clerk and six Los Angeles councilmen—all in all a cadre whose support for the notion that Eaton and Mulholland had agreed upon at the end of the previous year's trip was essential.

While it is not certain that Mulholland and Eaton began their journey in the same way, it seems only logical that they would have. On November 7, 1905, Eaton notes, the party—sans Mulholland on this occasion—traveled the hundred or so miles from Los Angeles to Mojave aboard the Southern Pacific's *Owl,* the night train leaving at 5:00 P.M. for Oakland and San Francisco. They spent the night in Mojave and set out the next morning in five wagons northward along the Owens Valley Stage Road, stopping for lunch at scenic Red Rock Canyon, later putting up camp for the night at Coyote Holes, having covered another forty-eight miles that day. They set out for Haiwee at 8:00 A.M. on the third day of the journey, taking eleven hours to cover forty-five miles of rugged canyon road. On Friday, the fourth day, they were through the pass and on to Olancha above Owens Lake by 10:00 A.M., and by 6:00 that evening they had covered another thirty-four miles to Lone Pine.

On Saturday morning, the party set out from Lone Pine at 7:00 A.M., and by 1:00 P.M. they had traveled another nineteen miles to Independence, where for the first time they encountered the object of their journey, the Owens River, crossing it at a point where, according to Eaton's notes, the water was sixty feet wide and five feet deep. It was there, Eaton said, that they "rested the remainder of the day."

The latter part of their route, Eaton said, was "frequently crossed by creeks and irrigating ditches, carrying beautiful streams of mountain water." At dinner that night, Eaton described the conversation as "a resume of the trip this far . . . the sentiment being thoroughly unamious [*sic*] that the City of Los Angeles was, by its wise action . . . causing our city to enter upon a new and unheard of era of prosperity, by securing this adequate and inexhaustible supply of pure mountain water."

Eaton's enthusiasm could be accounted for in part by the time that it had taken him to convince anyone in authority of the validity of a plan that he had hatched as early as 1892 and perhaps even before. As part of a land-owning family with sizable holdings in Pasadena, Eaton had heard talk of the Owens Valley since his youth. It was described as a bounteous pastureland in summer, the place where Los Angeles ranchers had driven their herds when the Southland was baking and dry. A family friend, J. H. Campbell, recalled as a thirteen-year-old taking a trip with Fred Eaton and his pioneering father, Benjamin, an attorney and former district attorney, to the Owens Valley in 1880, a visit during which the elder Eaton "took measurements of water on all the streams," with an eye toward somehow transporting it to the family vineyards in Pasadena.

Later, in 1892, Eaton spent a summer in the Owens Valley, where he saw vast untapped potential in harnessing the typical flood runoff from Bishop Creek and other tributaries of the Owens River. The *Los Angeles Herald* reported on July 10 of that year that he had joined with three other partners from Los Angeles and the Owens Valley to form the Olancha Land and Irrigation Company in order to "acquire and improve desert land." As he would later tell reporters, he had gone to the Owens Valley "to study the water situation with a view to colonizing the valley, having heard a great deal about the magnificent water supply of that region." The more utopian aspect of his vision did not survive, Eaton said, for, "After studying the situation carefully, I was convinced that a colonization project was not practicable . . . and I abandoned the idea."

Still, he had seen for himself those "magnificent" waters, and the memory never left him. As early as July 1892, he described the immense possibilities of bringing the Owens River water through the ancient blockade at Haiwee Pass and down Nine-Mile Canyon as least as far as the area around China Lake. "I saw more water going to waste in the Owens River than is contained in all the streams and rivers of San Bernardino, San Diego and Los Angeles counties combined," he said. "These great resources must be developed in the near future. The cost of diverting the Owens River and conduction [of] its waters upon the lands south of Owens Lake will be trifling when it is considered that more than 250,000 acres of magnificent land [in the desert north of Mojave] will be opened up to settlement."

Indeed, as W. A. Chalfant, chronicler of early Owens Valley history observes, as much as 20,000 to 30,000 "inches" of water, or about half again as much as Mulholland hoped to find for his city, went to waste when spring runoff overflowed the stream banks and flooded the Upper Owens Valley. (An "inch" or "miner's inch" was a commonly employed term, generally agreed at the time to constitute a flow of one-fiftieth of a cubic foot per second, or about 13,000 gallons of water per day, when the average per capita consumption in Los Angeles was in excess of 150 gallons per day. Today, water supply is more commonly measured in terms of acre-feet, a unit equal to a flow of 326,000 gallons yearly, or about 893 gallons per day—Los Angeles Department of Water and Power estimates current per capita consumption at 105 gallons per day.) Though by late summer, the Owens River might be running nearly dry at its headwaters, if that wasted runoff water had been impounded during the spring, it would constitute a valuable resource.

The concept of a vast cascade of sparkling mountain water was one that Eaton could never get out of his head, and when drought struck Los Angeles before and during his term as mayor, he worked on the refinement of his plan for Los Angeles to make use of all that wasted water. "My idea was to organize a strong company which should develop the great water power of the streams which pour down from the high Sierras and then combine with the electric feature," he explained to a reporter for the *Los Angeles Express* in 1905, "bringing the water to the San Fernando Valley . . . and from the sale of the electricity and water I was satisfied the project would be an inviting one."

Though he had mentioned the possibilities to others before his conferences with Mulholland and city attorney William B. Mathews (Mulholland would also testify before a panel of inquiry that Eaton had been in his ear about the idea since the early 1890s), one federal government engineer spoke for most when he noted that the idea seemed about "as likely as the City of Washington tapping the Ohio River." But in 1899, while Eaton was still in office, a development took place that would re-energize Eaton, when Congress passed the legislation that would create the US Bureau of Reclamation (it was formally the Reclamation Service, a subdivision of the US Geological Survey until 1907), with the charge of "investigating the extent to which the arid regions of the United States can be redeemed by irrigation."

When Owens Valley citizens became aware of the possibilities created by the new Reclamation Service, they were quick to press their desire for a serious irrigation project upon their government. In April 1903, before the Reclamation Service was officially open for business, new director Haynes Newell contacted the service's West Coast Chief of Operations Joseph B. Lippincott to see if he thought the Owens River Valley was worth inspecting. As Chalfant observes, the stated purpose of any such survey was, by law, to determine the feasibility "for construction and maintenance of irrigation works for the storage, diversion and development of waters for the reclamation of arid and semi-arid lands."

Lippincott, who had been employed by the US Geological Survey since 1889 and was recognized as an expert in matters of hydrology about the region, had also been employed in various capacities as an engineering consultant for the City of Los Angeles, including appointment by then Mayor Eaton to the board that quizzed Mulholland prior to the city's acquisition of the private water company. In response to Newell's inquiry, Lippincott dispatched a young engineer named Jacob Clausen to look around the Owens Valley.

Clausen's opinion was formed almost immediately: the valley was in fact an ideal candidate for consideration of an irrigation project of the sort envisioned by the agency, and he recommended to Lippincott that all public lands in the valley that could be benefitted by such a project immediately be "withdrawn" or protected from any claim by private interests in order to prevent speculators from acquiring and driving up the price of lands that could be improved for the good of the general public.

The development of a Reclamation Service project would mean the end of Fred Eaton's plans for a joint public-private venture involving valley water rights; when he got wind of the service's interest, he hurried to the Owens Valley in April 1904 while Lippincott was inspecting a site at Long Valley where employee Clausen had recommended placing the principal storage dam for the service's proposed project. Though Lippincott and Eaton knew each other well, both would maintain to the end of their days that Eaton made no mention of what was brewing in his mind.

At the end of August, the two returned to Long Valley on an ostensible vacation expedition for a group that included Clausen. Clausen later contended that all the while Lippincott waxed fulsomely upon the government's plans for the project in the valley. "Lippincott and I were talking all the time and Eaton was listening to everything we had to say," Clausen said, "and this probably was what Lippincott wanted to be done."

Whether Eaton had been tipped off by Lippincott has never been established. However, no one has ever questioned the fact that shortly after Eaton's return to Los Angeles, he and William Mulholland were riding a buckboard through the desert, all the way to Independence.

It was there at a lonely ford in the shadows of the Sierra that Eaton delivered on Mulholland's request to "show me that water source you have." The banks might have been less than bursting at that time of year, but Eaton was a respected fellow engineer, not simply a city booster. He would have most certainly passed along the service's estimation that the Owens River on average carried more than ten times the average summer flow of the Los Angeles River. It was an amount that surpassed the Chief's most dizzying estimates of need by two and a half times.

One can only imagine what Mulholland thought and felt as he stood by the Owens River, listening yet again to Eaton's urgent narrative of possibility and the need for the city to act quickly, before the Reclamation Service took over the river forever. Mulholland may have gazed out over the vast expanse of scrub and pasture that surrounded him and felt a certain regret, or at least a sympathy for the ranchers and farmers and merchants who were counting on the federal government to bring prosperity to this isolated place, for the beauties of the wilderness, even its desolation, were not lost on him: "Some men look and see only sand and rock, stretching endlessly," he once wrote. "Others gaze on the desert scene and read a sermon in the sand, the cactus and the flowers. Silence everywhere—majestic, wonderful."

Mulholland was neither a steamrolling developer nor a hard-hearted bureaucrat, but an individual possessed of keen intelligence and wry humor, much favored by the press, the citizens of Los Angeles, and the men with whom he worked. To those who served under him, his advice was as simple as it was welcome: "When the whistle blows, shut and lock the office door, leaving all worries and shop troubles behind that locked door. Then go home and have a pleasant time with your family, remembering never to cross your bridges until you get to them." Mulholland was by all accounts a sensitive man. But he was also driven by the logic of the utilitarian credo—the greatest good for the greatest number—and by his loyalty to his adopted place.

He and Eaton had spent some time taking measurements and elevations along the way to Independence, and he would spend more time in the region surveying and evaluating. But ultimately, he would agree that the waters by which they stood would provide all that Los Angeles needed and then some, and that, furthermore, it would be possible to move them 235 miles to the city, using only gravity to send them there.

He had wrestled with an imponderable question for a number of years, and he had finally encountered the solution. It would not be easy to carry out the project, but he welcomed the challenge. His city was counting on him.

DOWNHILL ALL THE WAY

IT WAS AN ARTICLE OF FAITH AMONG READERS OF THE Donald Duck comics series in the mid-twentieth century that all knotty problems standing in the path of human ambition could be overcome by a consultation with the *Junior Woodchuck Manual*. Donald himself disdained the *Manual*, but his nephews, Huey, Dewey, and Louie, swore by it. They once saved Scrooge McDuck's fortune after the *Manual* guided them in the repair of a geological fault threatening to swallow the gajillionaire's underground vault. Another time they found the answer to an equally grave question: How could they run away from overbearing Uncle Donald and an insufferable existence in Duckburg to a life of pleasure on the beaches of Florida?

The *Manual* gave them the answer. As a map in that volume made clear, Duckburg lay at the very top of the chart, somewhere in a spot vaguely corresponding to the American Midwest. Florida, of course, was at the very bottom. Thus, the solution was obvious.

They could get on their bicycles and simply coast to Florida. It was downhill all the way.

Readers may be amused by duck ingenuousness, but, according to family history, it was essentially that very vision that had overtaken Fred Eaton on a day in 1893 when he stood atop Mount Whitney, above the Owens Valley, and glanced southward toward the city that he would one day serve as mayor. Why stop with the notion of taking Owens Valley water to the Mojave? It was indeed a long way from these parts to Los Angeles, he told his oldest daughter, Helen Louise, who had accompanied him on that day's climb, but it was downhill all the way.

The basic premise had never left him, and while Eaton often said that he had gone so far as to commission his own preliminary survey of an aqueduct to Los Angeles, he had never been able to convince anyone that the idea held merit. In the brook-no-obstacles Mulholland he had finally found his man. To Mulholland, the waters coursing the banks where he and Fred Eaton stood represented the solution to an insoluble problem. All that was needed was to connect one half of the equation with the other, and Mulholland believed he could do just that.

The waters of the Owens River flowed through the valley at more than 4,000 feet. Los Angeles was a city that hugged the sea. Of course there were a few mountain ranges and impassable canyons that intervened along that 250 miles or so of "downhill all the way," but as Mulholland and Eaton had lurched and heaved their way from Los Angeles to Independence, Mulholland's practical engineering mind had been at work, making use of crude barometer measurements to estimate elevations, using a quarter-century of water master's experience to estimate stream flows. (A French physicist by the name of Louis Paul Cailletet had developed a barometric altimeter in the late 1880s, but the instruments as we know them would not come into widespread use until the 1920s.) In the end, Mulholland was convinced that all physical obstacles could be overcome and that a practical route for an impossible aqueduct was possible.

Though he would work out the details of the plan in the weeks and months to come, Mulholland intuited during that first trip that Eaton's idea was not a phantasmagoria, and that by combining a system of storage reservoirs, tunnels (more than fifty miles of them, including a five-mile bore beneath Lake Elizabeth), inverted steel siphons (twelve miles of these, perhaps the most dramatic features of the proposal), and stretches of canal, lined channels, and nearly a hundred miles of covered culvert, he could move the river they were looking at to Los Angeles. He had a sketch of his plan ready shortly after he returned to the city.

In Mulholland's mind, the real problem lay in finding the money to secure the necessary water rights and rights-of-way, and in doing so before the Reclamation Service could block them. While Eaton and Mulholland agreed that the Owens River could save Los Angeles, there was one feature of the ex-mayor's vision that Mulholland did not share. Eaton had been trying to convince Mulholland that the project should be developed as a public-private partnership.

Eaton's idea was to raise the funds necessary for acquiring the water rights to the Owens River from a group of private investors. In turn, the private consortium would sell or lease half the water rights to the City of Los Angeles while retaining the rights to the other half as well as the right to develop hydroelectric power along the course of the river and its tributaries. The city could do as it liked regarding the acquisition of the water and the funding of the enormous aqueduct project, but meantime Eaton and his investors would tie up the water rights and continue with the development of lucrative power plants. Eaton understood that the huge aqueduct project would necessarily have to be a public undertaking, but he was also convinced that, having conceived of the possibility to begin with, he was entitled to a proper businessman's return on all activities connected to the venture.

Now that the practical-minded Mulholland had determined that an aqueduct project was feasible, Eaton wanted to return to the New York offices of Dillon & Hubbard, the attorneys who had facilitated the city's takeover of the water company, to proceed with the formation of a private consortium. Mulholland, however, was not sympathetic to any private involvement in the ownership of Owens Valley water. He had been caught between such forces in his early days as superintendent of the private water company, and his experiences since the city had taken over showed him that taxpayers were best served by public ownership of the utility. He was also politically astute enough to realize that any hint of a collusion with private interests would likely stop the project dead. Given the Reclamation Service's stated interest in the same water, the issue could well come down to a standoff between the federal government and the city. If private interests were seen to be in league with the city, politics might override all logic.

Though Eaton was not happy, he reluctantly agreed to Mulholland's terms as spelled out by city attorney William Mathews. Acting as the city's agent and receiving an appropriate commission, Eaton would proceed to buy up options on about fifty miles of lands along the Owens River to establish the necessary riparian rights, as well as the lands farther north in Long Valley that would be needed for a storage reservoir. It would be Mulholland's immediate task, meantime, to complete the survey work necessary for the refinement of construction plans and to convince the water commissioners to come up with $150,000 for the options. All this work would have to be done in secret, for if word of the city's intentions got out, speculation would cause the price of the Owens Valley lands to skyrocket. (Reclamation Service engineer Jacob Clausen had estimated the prevailing costs of the land for the 140-foot-high reservoir he proposed for Long Valley Reservoir at about $21.58 per acre.)

It is not clear who told J. B. Lippincott what Mulholland and Eaton had in mind regarding the Owens Valley, but on September 17, 1904, even before the pair took their legendary journey, Lippincott wrote a letter to his superior in the Reclamation Service regarding the city's intentions: "I find that they are looking towards the Owens River for a solution." It was enough to bring Haynes Newell all the way to Los Angeles for a meeting between himself, Lippincott, Mulholland, and Mathews. Clausen was clamoring for the Reclamation Service to undertake a reclamation project in the Owens Valley, Newell told Mulholland, and there was a petition on his desk signed by 400 Owens Valley residents demanding that his agency proceed with all possible dispatch. Just what exactly were the city's intentions? Newell asked.

Mulholland replied that the city had not yet formulated a firm intention, but any detailed information in the service's hands that could help in making that decision would be welcome. Given that Clausen's report was a public document, Lippincott replied, he would have a copy sent over. It might be presumed that this meeting could not have gone better for the city, though as Lippincott would later testify, it was made clear that the Reclamation Service would never agree to step aside in deference to the city unless any proposed aqueduct "was public owned from one end to another."

By February 10, 1905, Lippincott had followed up on the matter to his superior Newell. "There is the possibility of our not constructing the Owens Valley project, but of our stepping aside in favor of the city," he wrote, though he did suggest that the agency should attempt to dun the city for some of the work completed in surveys and core drilling to determine the feasibility of the dam site at Long Valley.

Though he would always contend that he had never conspired against the interests of Owens Valley residents and only followed the path of securing "the greatest good for the greatest number" in making all his policy decisions, Lippincott was at the same time collecting substantial consulting fees from the city ($2,400 for his part in the survey of Southern California water sources at a time when his annual salary from the Reclamation Service was $4,200). If this was not enough to cast him as a devil in some quarters, other evidences of his connections to Eaton would damn him forever in certain eyes.

In the spring of 1905, Lippincott wrote Eaton a letter asking for his help evaluating the merits of various applications for rights-of-way across public lands in the Long Valley area that had been filed with the service. According to Lippincott, no one knew the merits of the land in that area better than Eaton, and no one would be better qualified to determine whether these applications could potentially interfere with any irrigation project that could one day be undertaken by the service.

Eaton in turn used Lippincott's letter to obtain detailed maps of the proposed irrigation project, which served as a convenient guide in selecting the properties he was buying up for the city. In addition, it was said by Chalfant and others that Eaton bandied Lippincott's letter about to give landowners the impression that he was in fact representing the Reclamation Service. Some residents also complained that Eaton was not above dropping hints that those reluctant to sell would have their lands condemned by the federal government or would find their lands crossed off the list of those slated to receive the forthcoming irrigation waters. "Sell out or dry out," was the implication.

Eaton did not need to use subterfuge in many of his dealings, which were centered farther south in the valley near Independence, where the land was ill-suited for irrigation and owners were happy to learn of an eager buyer. Moreover, it is questionable whether Mulholland had any immediate intention of building a dam at Long Valley above Bishop, where fertile pasture and farming land abounded. His primary concern was to divert the water from its course at a narrow point in the valley just above Independence, and, for the time being at least, store the waters farther south along the route to Los Angeles. However, if the city gained control of any potential dam site crucial to irrigation in the upper Owens Valley—the most obvious site being that in the Long Valley area—then the service would be unable to proceed with its irrigation project.

Key to Eaton's efforts was a rancher named Thomas B. Rickey, whose holdings covered most of the area where Clausen had proposed that the Long Valley Dam be built and with whom Eaton had already secured an option regarding his hydroelectric plans. Rickey, who also held substantial properties farther south, near Big Pine, had tired of ranching and was negotiating with more than one potential buyer when Eaton came along. As Rickey explained to Eaton, he wasn't looking to retire altogether—he also had an interest in a company that wanted to develop a power plant on nearby Bishop Creek for distribution to mining companies in western Nevada and he simply wished to wash his hands of the ranching enterprise.

When the former mayor traveled back to Owens Valley in an attempt to renegotiate the purchase of Rickey's land at a more favorable price for the city, it is certainly possible that he produced the letter from Lippincott, suggesting that he had connections that just might be able to help Rickey along in his new business endeavor. In any event, the deal for all of the Rickey ranch lands—in Long Valley and near Big Pine as well—was quickly done. A gleeful Eaton wrote city attorney Mathews on March 23, 1905, that "after a week of Italian work," the key option was in hand.

However, the matter was far from settled. Eaton had already spent about $30,000 of his own money on options, and the city—with water commissioners uncertain as to the propriety of expending money outside the county and lacking voter endorsement—had been slow to provide additional funds for the process. Though he complained in a letter to Mathews on March 25 that "If the City had money, I could buy up the entire river in 60 days," it took more than two months for commissioners to appropriate the funds to secure the options on Rickey's land.

All the while, Eaton had stewed. As he told reporters, it still rankled him that he had agreed to let the project be developed wholly by the municipality. While he had conceded to Mulholland and Mathews, he complained, "This I disliked to do, for it would deprive me of what I believed to be a splendid opportunity to make money."

Thus, by the time the water commissioners voted the funds for the Rickey options, Eaton had come up with a modification to the deal. The city had little immediate interest in most of the Rickey Ranch pasturelands, he knew, but they were certainly anxious to keep those in the Long Valley out of the hands of the Reclamation Service. Since the city was not named in any of the option agreements, he was personally in possession of an option to purchase Rickey's holdings for $500,000. Eaton proposed to pass along to the city options for about half of the property, in addition to the property's water rights and an easement to build a reservoir in the Long Valley that was significantly smaller than the one Clausen had proposed, for the sum of $450,000 (he later negotiated the price down to $425,000). Eaton would keep the other half of the land and the 5,000 head of cattle and other animals on it, worth $10 to $12 apiece.

Eaton would later complain, "The result was an agreement that I would turn over to the city all the water rights I had acquired at the price I had paid for them, except that I retained the cattle which I had been compelled to take in making the deals . . . and mountain pasture land of no value except for grazing purposes." However, he had gained control of a sizable amount of grazing land, equipment, machinery, horses, mules, and a herd worth at least $50,000 for about $15,000 of his own money. "Should I desire to continue in the cattle business," he continued dourly to reporters, "it will be necessary for me to invest about $150,000 additional in the purchase of suitable farming land." On the other hand, he had become an overnight cattle rancher at a pretty reasonable price. In the end, he would also receive about $10,000 in commissions for his purchases on behalf of the City of Los Angeles, according to final figures released by the city auditor.

Water commissioners, with little recourse and time running out, agreed to the deal on June 5. Though Reclamation chief Newell doubted that his service could prevail in any eventual standoff with the city, he pressed the commissioners for a public statement of intention that he could use to mollify his own superiors, who were already fielding complaints from interests in the Owens Valley wondering when the service was going to get busy on the reclamation project. On July 12, the registrar of the local federal land office, S. W. Austin, wrote his superiors in Washington to complain about Eaton. The former Los Angeles mayor had been "representing himself as Lippincott's agent," Austin asserted, and had "secured options on land and water rights in Owens Valley to the value of about a million dollars."

"Said purchaser now owns all the patented land covered by the government reservoir site in Long Valley, and also riparian and other rights along the river for about 50 miles." Austin explained that sellers "were all generously inclined toward the project and believed Eaton to be the agent of the Reclamation Service," though

Austin did admit that Eaton's stated contention was that he was buying the lands to create a cattle ranch. However, by the time of his writing, Austin said, Eaton's attempts to buy property at Haiwee Pass, in the desolate area south of Owens Lake, made it clear that these purchases were being made on behalf of laying a water pipeline to Los Angeles.

Abandonment by the service of its irrigation project, Austin argued, would "make it appear that the expensive surveys and measurements of the past two years have been made in the interest of a band of Los Angeles speculators." A few days later, Austin followed up with a similar letter of complaint to President Roosevelt himself. While the city's intentions had become a matter of public record, and the outcry had begun, it seemed for opponents a bit too little too late. On July 28, 1905, Mulholland returned from a foray to the Haiwee district.

"The last spike is driven," he reported. "The options are all secured."

REMOVE EVERY SPECTER

FROM THE TIME OF HIS VISIT TO THE OWENS VALLEY WITH Eaton the previous fall, Mulholland had been hard at work. In the subsequent three months, he had retraced their route several times, checking elevations and the flow rates of tributaries of the Owens River to verify his initial conviction that his plan was sound. "When my shirt got dirty," he later told a reporter for the *Examiner,* "I'd come in for a change, but otherwise, I kept to the theme."

He also worked hard on the water commissioners to convince them that, though the concept might seem outlandish, it was in fact workable. He was staking his reputation on it. If the undertaking was unprecedented, it was by no means impossible. Furthermore, it represented the only possible way to skirt the looming impasse that the Geological Survey's Mendenhall had cited: "Going to a distant source for its water supply is not merely wise, but is absolutely necessary if the City's future growth is not to be at the expense of neighborhood communities."

If there seems a contradiction in the suggestion that going after the water of the Owens Valley was fine, while scooping up the water sources of nearby established communities in Southern California was predatory, Mulholland saw it otherwise. There was no burgeoning city in the Owens Valley with established pueblo rights to the Owens River waters—instead there existed merely a group of ranchers and farmers who saw the vague possibility of growth and development there. It made no sense to encroach upon the already developed agricultural areas in the San Gabriel Valley and the Coastal Plain to the south, he told commissioners, for those communities were well-established extensions of what was in essence a vast, interdependent organism. To Mulholland, what a visionary might see as possible in a largely undeveloped Owens Valley simply did not measure up to what was certain in Los Angeles: with the 400-cubic-feet-per-second flow that his aqueduct would provide, *2 million* people could one day live comfortably where the present 206,000 had virtually exhausted the present water supply.

Critics would later contend that Mulholland underestimated the number of citizens that the flow of the Los Angeles River could support, with the water department's own estimates raised to 250,000 in 1928, 300,000 in 1936, and finally to 500,000, but the superintendent was by this point fixated on the future needs of the city he championed. As he later wrote, he had to contend with Eaton virtually at "sword's point" in order to get the ex-mayor to drop his demands that he retain control of half of the Owens River water as well as rights to the power-generating capacity of the aqueduct.

In April 1905, Mulholland led a fact-finding tour back to the Owens Valley with Water Commissioners John Fay and J. M. Elliott in tow, along with Mayor Owen McAleer, Eaton, and city attorney William Mathews. After listening to Mulholland's carefully detailed construction plans and viewing the various tracts that Eaton had encumbered and those he had his sights on, the commissioners asked Mathews what he thought of the legality of the matter and of the practicality of floating a bond issue that would eventually repay the costs of the options, some $233,000 before it was done. Once Mathews gave his blessing, the commissioners agreed.

As for the next step, convincing voters to approve the project, Mulholland was not overly concerned. He felt in his heart that the project was necessary and in his engineer's mind that it was feasible. He was by now quite a popular public figure, far more trustworthy to voters than any politician in the region, indeed a man of the people.

On Saturday, July 29, 1905, the day after Mulholland's return from sewing up the final options in the Owens Valley, the *Los Angeles Times* broke the story of what was becoming one of the worst-kept secrets in the city's history. "Titanic Project to Give City a River," the front-page headline of Harrison Gray Otis's paper blared. "Options Secured on Forty Miles of River Frontage in Inyo County—Magnificent Stream to be Conveyed Down to the Southland in Conduit Two Hundred and Forty Miles Long—Stupendous Deal Closed."

A sidebar called the news "the most important movement for development in all the city's history." One writer described Mulholland in heroic terms, telling of his having returned "scorched and browned by the almost intolerable desert wind and sun" to announce that "the vexed water question has at last been solved." The new water supply was described as "immense and unfailing"; it would allow Los Angeles to "forge ahead by leaps and bounds and remove every specter of drought or doubt." All of it was going to cost about $23 million the paper said, to be funded by a series of bond issues that would be "asked of voters."

Elsewhere it was claimed, "The price paid for many of the ranches is three or four times what the owners ever expected them to sell for. Everybody in the valley has money, and everyone is happy." While much in the inflated front-page prose had its kernel of truth, the latter assertion would prove to go down in newspaper history along with lines such as the *Chicago Tribune*'s "Dewey Defeats Truman." Initially, there was little reaction in the valley despite Chalfant's later observation that the story was as much news in Inyo as it was in Los Angeles. The transactions that Eaton engineered were perfectly legal, after all, and the *Times* had gone so far as to applaud the role that the Bureau of Reclamation's Lippincott had played in the process.

Lippincott, the paper, said, "lent valuable assistance in getting title to land in Owens Valley." Furthermore, the story said, "It is through Mr. Lippincott that the water board secured its concessions from the Government," and it went on to assert that he provided three government engineers to help map the aqueduct's route, "all the way from Charley's Butte to the San Fernando Valley." (That "Butte," near the point ultimately chosen for the aqueduct's diversion point, was named for African-American cowboy Charley Taylor, who was killed in 1863 while defending a party of settlers under attack by Native Americans.) "Any other government engineer, not a resident of Los Angeles," the story applauded ingenuously, "undoubtedly would have gone ahead with nothing more than the mere reclamation of arid lands in view."

As later developments would bear out, it was the sort of approbation that Lippincott could have happily lived without, but it also bears out how little was thought of the manner of acquisition in Los Angeles. Mulholland, in fact, suggested that the commissioners ought to write a letter of thanks to the Reclamation Service for all its help. Because his employment as a consultant for the city occurred at the same time that he was drawing a salary from the federal government, Lippincott suffered considerable criticism from valley residents and from coworkers such as Jacob Clausen, who had met his future wife while working in the area and would take a personal interest in seeing the irrigation project move forward. Reclamation chief Haynes Newell was reluctant to censure Lippincott publicly for fear that the service as a whole would look bad. In the end, Lippincott was removed by his superiors from any involvement with the Owens Valley project in March 1906, and in July of that year he resigned his position with the bureau and went to work for the city.

Meanwhile, of much greater controversy in Los Angeles was the fact that the *Times* had violated an informal agreement among all the principal newspapers to keep the matter of the proposed aqueduct off their pages until the work of securing the necessary land options had been completed. Once Mulholland had given the green light, it had been agreed, then all the papers could break the news simultaneously. Thus, when Otis went forward with the story without informing the others of his intentions, William Randolph Hearst's proletarian-leaning *Examiner* reacted angrily, taking on an antagonistic role as Mulholland attempted to rally public support for the $1.5 million bond issue announced in order to pay for the options and purchases in the Owens Valley. The story's appearance also caught Fred Eaton off guard. He was still in the valley when word came of what the *Times* had printed; as he later told reporters, he had to beat a hasty retreat to San Francisco to avoid a beating or worse. "When I go back for my cattle," he told a reporter, "they will drown me in the river."

The *Examiner* complained that Mulholland had not produced a detailed plan of the planned aqueduct and said that a project such as he proposed could not be built for any less than $50 million, more than twice the superintendent's estimate. Given that the total bonded indebtedness of the city stood at $7 million at the time, even Mulholland's figures seemed formidable; the price tag insisted upon by the *Examiner* would have put the matter out of the realm of reason. Mulholland, however, was unfazed. On the Monday following the appearance of the *Times* story, he told a Municipal League banquet that he had "examined every foot" of the Owens River and showed attendees maps of the project he had prepared. With his characteristic wit, he told prospective voters, "If you don't get it now, you will never need it."

The *Examiner* was not through, however. On August 24, the paper ran a story about the syndicate—including competing newsman Harrison Gray Otis—that had purchased the vast Porter Ranch tract in the San Fernando Valley, lands that would increase astronomically in value once Owens River water arrived. No wonder the *Times* was such an ardent supporter of the project, the *Examiner* said. Otis and the other rapacious businessmen who ran the city stood to line their pockets with ill-gotten gains.

It is hard to know how much effect this vendetta could have had upon Mulholland's campaign, for on September 2, 1906, matters took an unexpected turn, when, at the invitation of the Los Angeles Chamber of Commerce, William Randolph Hearst came to town for a consultation on the matter. Following that meeting, Hearst appeared at the offices of the *Examiner* and informed his editor that henceforth the position of the paper would be to help the City of Los Angeles with its bond issue. After he had delivered this somewhat surprising declaration, Hearst wrote an editorial for the next day's edition, recapping some of his paper's earlier reservations but concluding that, so long as the promise of the city's water commissioners to have Mulholland's plans vetted by an independent group of advisors held, the *Examiner* was dropping its opposition.

Though some have opined that Hearst's change of heart was

prompted by his desire to find support in Los Angeles for his presidential aspirations, it seems equally plausible that he agreed with the larger business community with whose representatives he had just met. If building the aqueduct would allow Los Angeles to grow and prosper, then liberals and those who sold newspapers to them would benefit along with conservatives.

Whatever the reason for Hearst's shift, Mulholland proved correct in his suggestion to a reporter that the citizens of Los Angeles "have always been in the habit of taking my word." On September 7 the bond issue was approved by a vote of 10,787 to 755. Though Mulholland had been confident, even he was surprised at the final margin. He had predicted something on the order of a 6 to 1 victory in his home precinct—the vote there was nearly 30 to 1.

Though opposition in the Owens Valley would continue to mount, creating a fresh set of problems in the years to come, the immediate obstacle in Mulholland's way had been cleared. The purchases of the lands under option were assured, and he could move on to the design and funding of the building itself. Only then did the gravity of what he had gotten himself into seem to strike Mulholland. "I put in the most anxious months of my life during that period," he would tell a writer for the *Times,* admitting that even he had come to question the viability of such an unprecedented undertaking. But still, buoyed by his bedrock conviction that it was "the right thing to do," he threw himself into what would become seven years of hard work.

There were innumerable modifications made as engineers (actually employed by the hydrographic branch of the US Geological Survey with the city paying expenses) went to work in the field to lay out the route that Mulholland had originally intuited. Key to his sense of the viability of the project from the beginning was the geological fact that during the days of prerecorded history, the Owens River, draining some 2,810 square miles of watershed in the surrounding Sierra, Inyo, and White Mountains, ran originally at least as far as China Lake in the Mojave Desert, nearly half the distance to the bounds of the San Fernando Valley. Until the late Pleistocene Era, some 11,000 years or so before, glaciers in the mountains fed a lake so huge that it overflowed the valley at Haiwee Pass and pounded down the canyons into the Indian Wells Valley and beyond. Eventually came the recession of the glaciers, a diminution of the annual runoff, and the eventual formation of Owens Lake as the terminus of the river, a process likely aided by upthrust from earthquakes at the southern end of the valley along historic fault lines at the base of the Sierra and Inyo Mountains.

In essence, Mulholland would simply be restoring the original flow of the Owens River, although he would have to modify the rate of descent to get waters all the way to Los Angeles without the use of cost-prohibitive pumps. Instead of dumping out onto the plains a hundred miles south at China Lake (elevation 2,200 feet), the "new" Owens River would be diverted at Charley's Butte at a point about 3,900 feet above sea level, to empty eventually into a pair of reservoirs (elevation 1,200 feet) in the northern San Fernando Valley below the Newhall Pass, about 225 miles away. From that point, the diverted waters would basically follow the same course as the traditional flow of the Los Angeles River.

As simple as the basic "downhill all the way" plan might sound, there was a great deal more involved. Because the Chamber of Commerce had, presumably as a result of its discussions with William Randolph Hearst, pledged that the city would not expend any more money than was necessary to preserve their interests in the land "until they shall have secured the approval of the entire plans by disinterested experts of the highest character," Mulholland would have to modify his plans in consultation with the panel.

In the end, the detail that Mulholland eventually came up with involved digging a gradually descending canal for sixty miles along the valley floor from the intake point above Independence near Charley's Butte through the Alabama Hills, skirting the site of Owens Lake and culminating in a seven-mile-long reservoir at Haiwee, elevation 3,760 feet. The waters would thus be stored at the head of Nine-Mile Canyon, the impoundment acting as a hedge against seasonal variations in runoff and allowing for the natural sedimentation process to aid in clarifying the water.

From Haiwee, the water would enter a covered conduit and drop far more precipitously—about 400 feet in 15 miles or so—to a series of tunnels bored through the rugged escarpments at the south terminus of Rose Valley. Once through the tunnels, water would be carried through a series of eight huge steel pipe "siphons," some eleven feet in diameter, down and back up the sides of a series of steep canyons, the very names of which give some sense of their character: Five-Mile, Deadfoot, Nine-Mile, and Sand Canyon.

Though the drop within a canyon and the ensuing climb could be as great as 850 feet, as at Jawbone Canyon, the water would be forced down and back up by the simple fact that the ultimate exit point of the line at the San Fernando Reservoir would be about 2,400 feet lower than the point of entry at Haiwee. Simply put, "water seeks its level," and though the great pipes were initially called "siphons" and later "inverted siphons" (because they resembled traditional siphons turned upside down), no priming suction was needed. Currently preferred terms for such structures, according to the American Society of Civil Engineers, include "sag pipe, sag line, sag culvert and depressed pipe," but perhaps Mulholland can be forgiven: "Jawbone Depressed Pipe" just wouldn't have the same ring.

Below the twenty-five-mile stretch punctuated by that series of siphons would run a relatively placid section of some twenty miles of covered conduit along the Salt Wells Valley to the Red Rock Summit, at about 3,200 feet. For about twenty miles south of that point, the aqueduct would leave Inyo County and pass into Kern County, crossing some of the most rugged terrain in its path and requiring another series of tunnels and siphons, including the massive undertaking at Jawbone Canyon. Once past Jawbone and Pine Canyons, the aqueduct would make a relatively mild descent of forty miles or so in covered conduit and pipe across the Antelope Valley to the Los Angeles County line southwest of Mojave, and then another twenty-plus miles over the southern valley floor to the Fairmont Reservoir west of Lancaster, at the foot of the San Gabriel Mountains below Lake Elizabeth.

The elevation of the Fairmont Reservoir would lie at about 3,000 feet, and with the reservoirs at the planned terminus at less than 1,300 feet, the drop over the aqueduct's last thirty-two miles or so would be by far the steepest, just about twice as great as the water's descent over the 190 miles to that point, from an average of about five feet per mile to more than fifty feet per mile. As anyone who has navigated the Newhall Pass and the twisting roads of San Francisquito Canyon knows, the terrain of that thirty-two-mile stretch is anything but gentle. To cross those lands would require another daunting series of tunnels and siphons crossing Soledad Canyon, Deadman Canyon, and more. Most ambitious of all the links in the aqueduct's route was the five-mile-long tunnel that would have to be bored through the solid granite of the San Gabriel Mountains beneath Lake Elizabeth.

The notion of the Elizabeth Tunnel constituted a significant change in plans for Mulholland, for he had originally conceived of a pair of six-mile-long tunnels farther east in the mountains near Acton. Daunting as the Elizabeth Tunnel undertaking was, however, it would shorten the length of the aqueduct by twenty miles.

In all, the final plan called for a system 225.87 miles long, from Charley's Butte to the San Fernando Reservoir, with roughly 22 miles of unlined conduit or canal, 164 miles of concrete lined or covered conduit, 28 miles of bored tunnels, and 12 miles of steel pipe and siphons. And it could all be completed for $24.5 million dollars, Mulholland vowed, including the $1.5 million already expended for the land.

By way of comparison, plans were going forward in New York for construction of the Catskill Aqueduct, a 163-mile-long addition to the Croton System that had been in operation since the 1840s (and rebuilt in 1890). Construction on the Catskill project was to commence in 1907 (it would be largely completed in 1916, by which time about $140 million had been spent) and would end in toto at $177 million, principally owing to costs in acquiring the necessary right-of-way. The Catskill Aqueduct was a vast undertaking, to be sure—one of the two grand water projects of the time, but in terms of size and the daunting nature of the terrain to be traversed, Mulholland's task was positively monumental and questions of budget were crucial.

In addition to the cost-saving change regarding the Elizabeth Tunnel, Mulholland had also been forced by the consulting panel to abandon any thought of an impoundment dam at Long Valley. There would therefore be no way of regulating the seasonal vagaries of the Owens River above the diversion point at Charley's Butte, but the savings—Reclamation Service engineer Clausen had pegged the cost of such a dam at $750,000—demanded the cut. At the same time that Mulholland was refining his vast plan, he was also working to secure the necessary grants of right-of-way for any federal lands that would be necessary for the project to cross. California Senator Frank Flint had supported the concept from the beginning and in the summer of 1906 he introduced a bill that would allow for the acquisition of any public lands necessary for the project at $1.25 an acre. The bill sailed through the Senate, but when it reached the House, Representative Sylvester Smith of Inyo County, a member of the Public Lands Committee, proposed an amendment demanding that the residents of the Owens Valley retain rights to the river waters. According to Smith's proposal, only the water left over after the valley's irrigation needs were met could be diverted to Los Angeles—and only for domestic use, not for irrigation.

The city itself had stated that it could possibly require an additional supply of 2,500 "inches" of water in twenty years, Representative Smith claimed (Mulholland had in fact suggested about three times as much). But the aqueduct was being designed to carry 20,000 inches. Surely an allotment to the city of 10,000 inches would be sufficient, he said, leaving the Owens Valley with more than enough water for the plans of the Reclamation Service. Debate ensued.

Mulholland traveled to Washington to confer with Senator Flint and suggested that perhaps the city could live with Smith's compromise. Flint, however, thought that Gifford Pinchot, chief of the Forest Service and a close friend of President Roosevelt's, might be willing to intercede on the city's behalf, making compromise unnecessary. At a meeting on the evening of June 23, Roosevelt listened as Flint, Pinchot, and Charles Walcott, director of the US Geological Survey, held forth on the matter, which was by then before the House Committee on Public Lands.

Among other issues, Roosevelt's visitors stressed that the population of all of Inyo County—of which the Owens Valley constituted only a part—was less than 3,000 in 1880 and stood at 4,377 in 1900. During that same period Los Angeles had grown from 11,000 to more than 102,000, and in 1906, Los Angeles County had a population of more than 350,000. The valuation of all property in Inyo

County for 1906 was $2.6 million. The valuation of property in Los Angeles alone that year was $203 million. There were somewhere between 30,000 and 40,000 acres of land being irrigated in the Owens Valley at the time, principally in alfalfa grown to feed livestock, and while Jacob Clausen estimated that a reclamation project could bring more than 106,000 additional acres under irrigation, it was the city's estimate that no more than 60,000 to 80,000 could be feasibly added.

Though Roosevelt would enjoy an enduring reputation as a trust-busting friend of the people and the environment, he was also a Republican, an expansionist, and was motivated above all to encourage development of a vital national economy. When all had finished, the president found himself looking on the one hand at the prospects of bringing 80,000 to 100,000 acres of far-flung land under irrigation versus the future needs of a city of 250,000 that seemed destined to grow to immense size. It did not take him long to decide. Two days later, the president called in Ethan Hitchcock, secretary of the interior, and in the presence of Hitchcock, Flint, Walcott, and Pinchot, dictated a letter so that there could be, as Roosevelt put it, "a record of our attitude in the Los Angeles Water Supply Bill."

Roosevelt said that he understood Hitchcock's concerns that without Representative Smith's amendment it was feared that Los Angeles might use the water in excess of its current needs "for some irrigation scheme." However, Roosevelt, said, he was convinced by Senator Flint's counterargument that while there could indeed be some present surplus in the waters obtained from the Owens River, the city was seeking to ensure a sufficient water supply for a half century to come; it would therefore have to draw the full amount of water from the very outset or risk its being absorbed by other interests in the meantime.

"It is a hundred or a thousand fold more important to the state and more valuable to the people as a whole if used by the City than if used by the people of Owens Valley," Roosevelt said, echoing Flint's urgings. In referring to a subplot that had complicated matters all along, Roosevelt also dismissed the objections of various electrical power companies, including the formidable Edison interests in Los Angeles, who feared that the city would develop the power resources attendant to the streams in the Owens Valley and along the route of the aqueduct, depriving them of future business opportunities.

"The people at the head of these power companies are doubtless respectable citizens," Roosevelt said, and he agreed that there was no law against their seeking their own pecuniary advantage. "Nevertheless," the president continued, "their opposition seems to me to afford one of the strongest arguments for passing the [Flint] law, inasmuch as it ought not to be within the power of private individuals to control such a necessary of life as against the municipality itself."

Thus, Roosevelt concluded, he was asking that Flint's bill be approved without the encumbrance of the Smith amendment. In an apparent slap against the interests of the San Fernando Valley syndicate, Roosevelt did request that the bill include a prohibition against the city's ever selling or leasing to any corporation or individual—except for a fellow municipality—the rights to Owens Valley water for irrigation purposes. He acknowledged that the interests of the "few settlers in Owens Valley" were genuine, but that interest "must unfortunately be disregarded in view of the infinitely greater interest to be served by putting the water in Los Angeles."

When he was finished, Roosevelt glanced about the room, then felt compelled to add a postscript. "Having read the above aloud," he said, "I now find that everybody agrees to it—you Mr. Secretary, as well as Senator Flint, Director Walcott and Mr. Pinchot, and therefore I submit it with a far more satisfied heart than when I started to dictate."

For the Owens Valley, being told by the president that its interests were insignificant would likely have seemed as much a personal blow as an economic one. (Local residents are still rankled by Roosevelt's order of May 25, 1907, creating the Inyo National Forest out of the largely treeless public land in the region in order to facilitate the withdrawal of public lands necessary for the aqueduct right-of-way.)

But for Mulholland, Roosevelt's stance was not only a justification but also one more necessary step toward making the aqueduct possible. Eaton's purchases above the intake point provided in the aggregate about 15,000 inches of water rights, based upon the appropriations that were the property of the previous owners. In addition, the city had acquired virtually the entirety of the fifty miles of largely scrub land bracketing the river banks from the intake point southward to Owens Lake, which would provide another 5,000 inches of runoff water by Mulholland's estimate.

With the rights to the water secured and the federal rights-of-way assured, all that remained was to secure the passage of the bond issue for construction. And of course there was also the small matter of doing the work itself.

HAVE WATER OR QUIT GROWING

When William Mulholland was introduced to share the details of his project at the Municipal League banquet on August 15, 1906, reporters noted that it took several minutes for the applause to die down before the superintendent could speak. "What can I say of him except that he is Will Mulholland," toastmaster E. H. Lee began, before a chorus of cheers erupted. Lee finally gave up further attempts at introduction and called Mulholland to the podium.

Mulholland quieted the crowd and began his first public appeal with a declaration: "We must have the Owens Valley water," he said. "The chance to acquire such a supply is the greatest opportunity ever presented to Los Angeles." He touted the runoff from the Sierra as more pure than that of the Los Angeles River, "clear, colorless, and attractive," and containing only a third of the minerals of the present city supply. Though there was some sediment and pollution introduced from existing farming and irrigation practices in the Owens Valley, Mulholland explained that sedimentation and natural bacterial action during storage in the Haiwee and Fairmont Reservoirs planned along the route would remove any such contaminants. Once the water left its second storage point at Fairmont, it would remain in covered pipes and conduit until it emerged from a customer's tap.

While the subject of water treatment is beyond the scope of this book, it may be worth pointing out that chlorination of municipal water supplies was a practice nonexistent in the United States at the time (the first such application took place in New Jersey in 1908), and Louis Pasteur's so-called germ theory of disease transmission was a concept still in its infancy. It was not until 1914 that the US Public Health Service published standards for bacteriological quality of drinking water. Meantime, scientists had begun to realize that suspended particulates in water—including fecal matter—could house harmful pathogens, including typhoid, dysentery, and cholera, so that efforts had begun to decrease "turbidity" in drinking water supplies, allowing particulates to settle out. But while there were a very few sand-filtration installations in the East, the accepted standard for water "treatment" at the time was a period of storage in reservoirs, where sunlight, oxygenation, and other natural processes could assist in the purification process.

Following Mulholland's discourse on the purity of Owens River water (he had been drinking plenty of it, he reminded everyone), he admitted to a questioner that the project was indeed a big one. But if he harbored any serious qualms, he was certainly not going to admit them publicly. This project was in the end a simple one, he assured the audience of some 175 movers and shakers (oil tycoon E. L. Doheny, after whom the state's first public beach would be named, was present, as was developer I. N. Van Nuys).

Though it is uncertain just how familiar Mulholland would

have been with it, another unprecedented engineering project was just getting underway on the other side of the continent: self-made Florida railroad titan Henry Flagler's effort to build an "impossible" rail line across 153 miles of largely open ocean from Miami to Key West. Faced with similar questions from those who wondered how such a thing could be accomplished, Flagler famously responded that it was simple: "First you build one arch, and then another. And before you know it, you're in Key West."

Likewise, Mulholland had his own ready explanation: "The man who has made one brick can make two bricks," the superintendent said. "That is the bigness of this engineering problem. It is big, but it is simply big."

Mulholland organized the "bigness" into separate divisions, of which there would ultimately be eleven: the first, the Owens Valley, was to be concerned with digging the canal from the diversion point near Charley's Butte to the Haiwee Reservoir; the reservoir itself was to be the work of the Olancha Division; the next was charged with the construction of the conduit down the Rose Valley from Haiwee; then came the work through the badlands below Little Lake; the Grapevine would handle the difficult building of the tunnels and siphons that would carry the project through that rugged terrain to Salt Wells Valley, where the Freeman Division would be in charge of the relatively straightforward work on eighteen miles of gently sloping covered conduit; the Jawbone Division would take over from Red Rock Summit, where another series of daunting tunnels and conduits were required, including the eponymous Jawbone Siphon.

The Mojave and Antelope Divisions would divide the task of carrying the line across the Antelope Valley to Fairmont Reservoir, which was designed to ensure an uninterrupted supply of water in case of any breaks in the line above (though it would also be capable of providing a steady flow to a power plant that Mulholland suggested could be added in San Francisquito Canyon, taking advantage of the great drop in elevation through the San Gabriel Mountains). The Elizabeth Division's responsibility would be the digging of the great tunnel through the San Gabriel Mountains beneath Lake Elizabeth, and the Saugus Division would carry the water on through the San Gabriels into the San Fernando Valley, including the boring of a formidable tunnel north of the Newhall Pass.

In addition to the construction work itself, Mulholland was also faced with the question of how to bring supplies to the line. Train service was in place as far as Mojave, but north of that settlement there existed only the rugged stagecoach and wagon track that Mulholland and Eaton had used on their initial foray. The building of a railroad between Mojave and the Owens Valley would be a boon to all, Mulholland said. At present, any freight out of the valley had to be carried by what he termed "very defective transportation facilities," a limited system of narrow-gauge rail lines that made a circuitous connection of some 600 miles to San Francisco, crossing passes as high as 7,000 feet. In Mulholland's opinion, this was the actual obstacle that had blocked development in Inyo County and the Owens Valley. A proper railroad running southward out of the valley would be only 230 miles long, with a downhill run to Los Angeles most of the way.

There was also particular need of a spur line to the area south of the Red Rock Summit where the massive sections of steel pipe for the Jawbone and Pine Canyon Siphons would have to be transported, but Mulholland understood that it was unlikely that the city would be willing or able to put itself in the railroad business. He was already negotiating with the Santa Fe and Southern Pacific companies to build that supply line as well as to provide favorable freight rates to the city for carrying materials to the work camps.

One ancillary business would have to be begun by the city in order to build the line, Mulholland said, referring to the necessary construction of a cement plant somewhere along the route. Accordingly, he identified the 3,000-acre Cuddleback Ranch near Mojave as a site containing sufficient deposits of lime and clay for making cement. The city would have to purchase that property and build a mill, which would cost about $300,000, but given the amount of cement they would need—about 1.3 million barrels—along with the savings in transportation costs by having supplies virtually on site, even that additional outlay would ultimately result in savings. He had already had test batches of cement made from materials taken from the property, he told commissioners in his proposal, and cement expert Edwin Duryea, hired on from the Reclamation Service for the purpose, had pronounced the product "superior."

All of Mulholland's preparations were eventually compiled in a document entitled "First Annual Report of the Chief Engineer of the Los Angeles Aqueduct" that was presented to a new city agency, the Board of Public Works, created on March 1, 1906, in accordance with a revision to the city's charter. The new board was, in Mulholland's words, to take charge "of all expenditures of money derived from the sale of municipal bonds." Since the Los Angeles Aqueduct would be funded by bonds, it would become the official body overseeing the project, and all positions on the aqueduct would be filled under Civil Service Commission regulations and approved by the Board of Public Works.

During the construction process, Mulholland would essentially wear two hats. He would continue on as chief engineer and manager of the City Water Department, reporting to the Board of Water Commissioners; but he would also function as chief engineer for the aqueduct, reporting in that capacity to the Board of Public Works. To ensure that all went smoothly, an Advisory Committee was formed, composed of the presidents of the Water and Public

Works Boards, Mulholland, his chief assistant for the aqueduct, J. B. Lippincott, and city attorney William Mathews.

By March 31, 1907, when Mulholland presented his first report to the new board, all was up and running. Three surveying parties were in the field, a commissary was already in place at Mojave, and a construction camp was established in the Owens Valley, with a power shovel at work digging the main canal.

An April 1 report by Aqueduct Disbursing Officer W. M. Nelson showed that $1,294,251 of the $1,512,246 realized in the initial bond sale had been spent, leaving only about $243,000 in the kitty. What the city had to show for its expenditures consisted primarily of 76,581 acres of land. That outlay totaled $1,035,073.77, including $13,014.90 in commissions, $10,666 of which went to Fred Eaton.

Only about $160,000 of the total had been expended for actual construction on the aqueduct itself, the report stated, including the digging and grading of five miles of the main canal in the Owens Valley and six miles of wagon road. The rest had gone for the completion of some 700 miles of surveys and studies of route conditions, the purchase of an undeclared amount of right-of-way, machinery, equipment, and tools, along with, in one reporter's words, "100 other things leading up to active work on the construction of the aqueduct as soon as voters make their decision on the bond issue Wednesday, June 12."

The latter was the last real impediment to full-scale work on the project, though there was not much concerted resistance standing in the way. As a *Los Angeles Herald* story (April 1, 1907) released in tandem with news of the disbursement report opined: "The past week opened one of the most remarkable campaigns for one of the most remarkable projects that any city, ancient or modern, has ever undertaken." A committee formed by the Chamber of Commerce, the Merchants and Manufacturers Association, and the Municipal League had opened an office and begun to circulate calls for every good citizen to turn out to vote in favor of the bond issue so that construction could begin in earnest. Democrats, Republicans, and even the unaffiliated were allied in support of the measure, the report claimed, along with labor and capital, prohibition and antiprohibition organizations, and the city's "five great newspapers," presumably the *Herald*, the *Times*, the *Examiner*, the *Record*, and the *Evening Express*.

Nor were supporters of the aqueduct asking voters to lend support to a completely unfamiliar undertaking. Any schoolchild would have heard at least passing mention of the thirteen aqueducts of ancient Rome, a 500-mile system largely credited with turning that city into civilization's center. The largest of those, the Aqua Marcia, measured a "mere" fifty-six miles in length, delivering about 50 million gallons a day. Compared to what Mulholland was proposing, the Aqua Marcia was little more than a section of garden hose.

To help galvanize support, the joint committee of backers prepared a campaign pamphlet based upon Mulholland's report to the Board of Public Works. Among other things, the "Owens Valley Water Primer" reiterated that there was no other sufficient supply of water available to ensure the city's continued growth, which unless augmented would be exhausted in "five, or six years." In countering some claims by those who had described the waters of Owens Lake as undrinkable, the primer pointed out that the waters of the river were being diverted at a point fifty miles upstream from the lake, where they were pristine. Samples drawn at that point had been tested by the federal government, the primer stated, as well as by three other agencies, all of whom had reported favorably.

As to the feasibility of the project to which they were being asked to commit so many of their tax dollars, taxpayers were assured that experts were in agreement. It would take five years to complete the construction, at a cost of about $21 per foot. Furthermore, in a nod to Mulholland's popularity, voters were reminded that there was a firm hand at the controls of the project. Since the water company had been acquired by the city, rates had been reduced by 10 percent and were less than half the rates charged by those supplying San Francisco, Oakland, Alameda, and Berkeley. In five years, the City of Los Angeles had greatly enlarged its system and in 1906 earned net revenues of $652,416. The aqueduct, while expensive, would eventually pay for itself, just as the city's system was flourishing.

As for the power question, Mulholland had been downplaying the matter. He pointed out that there was no mention of paying for power plants in the bond issue—the people could determine that matter at some later date. The prospectus did point out that while the cost of building power plants and transmission lines from San Francisquito Canyon to the city would be about $4.5 million, it was a potential add-on that would easily pay for itself. In 1906, the city had paid $156,000 to private companies for its street lighting alone.

The primer also reminded voters that all of the bonds would not be sold immediately, but would instead be parceled out as the work on the aqueduct progressed, thereby saving on the costs of servicing the bonds' debt. There would be no payment on the principal until the project was complete, and the cost to the average homeowner in the interim would amount to about $2.66 per year, based upon an assessment of $1,600 for a $3,000 home, the operating standard of the day. Supporters of the bond issue also stressed that 90 percent of the monies to be expended on the project—with the exception of the purchase of the steel pipe and the heavy machinery to be contracted for back East—would circulate through the local economy.

Already, the document stated, there were upward of 120 local men employed in the undertaking.

The primer might have proved informative for some, but as might have been expected, it also became a convenient target for opponents. Soon a parody was circulating, aping the question-and-answer format devised by the Chamber of Commerce. While the original had answered the question *"What is the proposed Los Angeles Aqueduct?"* with a straightforward rundown of statistics, the parody described it otherwise: *A: "It is a piece of gigantic folly that will cost the taxpayer fifty millions of dollars or more."*

One writer for *Out West* summarized the various objections that popped up, some of which had circulated since the days of the initial bond offering campaign: the project was patently impossible, and even if it *were* possible, it would take twenty years and at least $50 million to build. There wasn't really as much water in the Owens River as Mulholland said there was, and even if there was, most of it would evaporate before it ever made it all the way to Los Angeles. Opponents argued essentially that only the mind of a duck could conceive that this phantasmagorical water could flow downhill from the stipulated location—it would necessarily have to be pumped over the many mountains and ridgelines that intervened. Furthermore, there would have to be two lines built in case one should be destroyed by the earthquakes common to the territory, and if that were not possible, at least the one line should be built out of sturdy steel pipe.

Though the private power companies were vigorously opposed to the bond issue and contributed on the sly to opposition efforts, they kept quiet publicly because their self-interest would hardly have been persuasive to voters. (The companies claimed to be opposed to the prospect of the city's distributing power more than generating it, arguing that the expense of erecting a duplicate distribution system was a waste of taxpayers' money.)

Mulholland wisely took advantage of the silence of the companies, declaring in his First Report, "the power situation is considered as wholly independent of the proposition of supplying water, and should stand on its own merits." To underscore his position, Mulholland had hired consultant Ezra Scattergood as chief electrical engineer for the aqueduct project. The project would very likely need its own generating facilities to power operations in the distant desert, but all things electrical would be under Scattergood's purview. By the end of May, Los Angeles newspapers reported that both the Los Angeles Gas and Electric Company and Edison Electric had endorsed the proposed bond issue, and with that the matter of the bonds' passage was essentially settled.

On Wednesday, June 12, 1907, more than 24,000 voters went to the polls and approved the measure in every precinct. The final tally showed nearly 22,000 in favor and slightly over 2,000 opposed, a margin of almost 11 to 1. It was time, it seemed, to go to work in earnest.

BRICK UPON BRICK

"I WANTED THIS FIGHT," MULHOLLAND WOULD LATER TELL A reporter, referring to the massive undertaking that ensued. "When I saw it staring me in the face, I couldn't back away from it. . . . I didn't want to buckle down and have to admit that I was afraid of the thing, because I never have been—not for a second."

Yet Mulholland would find that more obstacles still stood in his way. Though city attorney Mathews quickly set off to New York City to sell the new issue to what he hoped were eager investment firms, he would have known that he was facing a difficult task. The stock market was in the midst of a slide that had begun in late 1906 when speculation in railroad securities began to drop.

By the time Mathews reached New York, the market was down by 20 percent overall, and a number of banks, as well as the New York Stock Exchange itself, were teetering on the edge of collapse. The "Panic of 1907" was finally halted by the storied intervention of steel magnate and investor J. P. Morgan, and would ultimately lead to the formation of the Federal Reserve System in 1913. But the effects of the crisis were profound, and Mathews was soon cabling that it was "impossible to dispose of the bonds" at their current rate. He advised that the City Commission authorize an increase from 4 percent to 4½ percent, an action that they took immediately, to negligible effect.

Yet Mulholland needed money and he needed it right away. Suppliers would wait only so long to be paid, and if he wasn't able to meet his payroll, workers would simply walk off the job. Appeals to local banks, insurance companies, and private investors resulted in little investment. The effects of the downturn had spread westward, and the city was even having trouble paying its police and firemen.

There was some good news during the period, for Mulholland was able to strike a deal with the Southern Pacific Railroad to build a rail line between Mojave and the Owens Valley as well as to extend preferred rates for carrying supplies along the construction line. The same economic downturn that closed any number of mining operations across the West also meant that Mulholland would have little trouble recruiting an experienced workforce for his digging and tunneling operations, so long as he could find a way to pay them. He was eventually able to secure waivers exempting most laboring classifications from Civil Service requirements, freeing him to hire men whose skill with picks, shovels, mule teams, and blasting caps far exceeded their abilities as test takers.

Even with the financial constraints placed upon him, by year's end Mulholland had 327 men at work on the project. There were twenty miles of road constructed in the area around Haiwee Reservoir, and work was underway at the North and South Portals of the Elizabeth Tunnel, assumed to be the most difficult of all the sections of the work. Mulholland had also negotiated a favorable contract for the 800 miles of copper wire he would need to string electrical supply and telephone lines along the construction route. Still, as J. B. Lippincott informed Mulholland, the last of the monies from the 1905 bond issue had been expended, and the aqueduct project was overdrawn. Though city commissioners voted grudgingly to transfer $10,000 from the water department's general fund to the aqueduct account, it was clear that a work stoppage loomed. And then fate intervened on December 27, when the State of California announced that it would buy $510,000 worth of bonds, a gesture of faith in the project and one calculated to aid the ailing economy.

Though only a drop in the proverbial bucket, the state purchase was the first major infusion of cash to the project in more than two years, and it unlocked something of a flow. Over the first half of 1908 came further purchases from local institutions and investors that brought total bond sales to a bit more than $1 million, allowing Mulholland to pay the backload of debtors' claims and begin operations in earnest.

By the middle of 1908 there were more than a thousand men at work, though most were involved in tasks related to providing the necessary infrastructure before the building of the aqueduct itself could begin: improvement of roads, the laying of power, phone, and water supply lines; completion of the cement plant at Monolith, sixteen miles west of Mojave; building generating facilities in the Owens Valley sufficient to power operations up and down the aqueduct; assisting the Southern Pacific in the completion of the rail line; the building of work camps, cookhouses, machine shops, barns, and corrals; and the assembly of heavy equipment, steel plate, timber, coal, and livestock from hundreds, even thousands of miles away.

All of this was taking place in a region where almost nothing had existed before, a vast, often blistering terrain where yearly rainfall averaged less than six inches and temperatures rose to 120 degrees in the summer and fell below zero in the winter. At the time Mulholland began his work, it could take as many as five days for a letter to be carried from Bishop to Los Angeles. It was an area that even the Native American populations had deserted centuries before. And now he was leading a modern army of construction trying to do what nature herself seemed to have decided against: providing a water supply to a desert place along the Southern California coast.

One reporter, sent out for a look at what Mulholland had managed, gazed from a cliff top out over the landscape of the Jawbone Division and marveled "as much at the ingenuity" of the engineers he saw at work, "as at the effrontery of Mulholland in daring to return from his trip of inspection and report to the Water Board that the thing was possible of accomplishment." Furthermore, it was nearing the middle of 1908, and for all the impressive strides Mulholland had made, the money had nearly run out again.

A payroll for a thousand men loomed due on July 1, and there was another $1 million due to various creditors. In the meantime, there had been a new chairman appointed to the Board of Public Works, Adna R. Chaffee, a retired general and longtime friend of Harry Chandler, son-in-law of *Los Angeles Times* publisher Harrison Gray Otis (and whom Otis had appointed general manager of the paper). Chaffee had accompanied Mulholland on a March 9, 1908, inspection of the project reported on by the *Times,* which included a stop at the daunting tunnel at Lake Elizabeth. Things were going well there, with almost a thousand feet of granite excavated by hand in the three months the work had been underway. Based on that progress, and with electrical power soon to reach the tunneling site, Mulholland told Chaffee that this crucial part of the project could well be completed ahead of schedule and nearly $700,000 under budget.

At the time Mulholland was delivering this news to Chaffee,

the inspection party was standing at the end of a five-mile city-built road off the main grade through the San Gabriels, outside the entrance to the tunnel at about 2,900 feet of altitude, with the crest looming another 1,000 feet above them. Mulholland bent to pick up a fragment of rock that had been recently chiseled out. "We're lucky," he told the group. "This is gneiss rock. I don't think we're going to hit any real granite in there."

Mayor Arthur Harper, who was along for the trip, took the fragment from Mulholland and hefted it. "Very nice rock," he repeated, clearly bewildered.

Mulholland smiled. "That's *g-n-e-i-s-s*, Mr. Mayor." The superintendent went on to explain that gneiss was a striated formation, much easier to cut through than solid granite, which is essentially solidified lava.

Chaffee was duly impressed with Mulholland's projections, but his mind was on something else. He had noticed a miner with his foot wrapped in a towel, the victim of an accident in the tunnel earlier that day. As former chief of staff of the US Army, Chaffee was attuned to personnel matters related to demanding engineering projects.

If the City of Los Angeles was going to put hundreds or perhaps thousands of men to work under difficult conditions, he believed there should be doctors in place to look after them. One of the commissioners piped up in agreement: perhaps they could station a doctor in Mojave and provide him with a motorcycle with which to ride between the camps, checking on the men.

But "to depend on a fellow with a motorcycle when human limbs and lives are in the balance" was hardly what Chaffee had in mind. There would have to be a physician and proper treatment facilities on-site. Accordingly, Chaffee asked that Mulholland, in addition to everything else he was faced with, see to the establishment of a Medical Department for the project, with medical tents placed at every encampment and permanent hospital buildings erected at each Division Headquarters.

The city would provide the tents and buildings to this Medical Department, Chaffee decreed, as well as offer transportation and basic support services such as water, power, and phone. However, the responsibility of providing the medical care and staffing was contracted to former Los Angeles County Hospital Superintendent Dr. Raymond J. Taylor. Taylor would in time come to oversee medical care for more than 10,000 men who worked for the project. In addition to supervising a medical staff and the treatment of sick and injured workers, his department was also responsible for oversight of their diet and nutrition, sanitation, disease prevention, mental health counseling, and more.

The contract called for workers making more than $40 a month to have $1 withheld from their wages, and those making less than $40, 50 cents, these sums diverted directly to Taylor and his partners. In turn, the men would be assured medical, hospital, and surgical coverage, "except for venereal disease, intemperance, vicious habits, injuries received in fights, or chronic diseases acquired before employment."

While there had been numerous suggestions that the city subcontract virtually all the work on the aqueduct, Mulholland had insisted from the beginning that he would be far better able to manage the costs of the undertaking by overseeing his own workforce. Ultimately, save for a short section of conduit in the Antelope Valley, the contract let to Taylor for the Medical Department and another to a man named Joe Desmond for commissary services constituted the only significant work on the entire line not done by Mulholland and his men.

According to Taylor, the arrangement between himself and the city ran smoothly, though his opinion that Desmond, just twenty-eight at the time, was more of a promoter than a commissary man would prove to be accurate. At the outset, Taylor repeated an observation regarding his operations that could have been the refrain for the project as a whole: "This was more or less of a gamble on our part as none of us knew very much, nor could we learn very much, about this sort of thing." The closest such undertakings had been army campaigns and the forging of railroad extensions through wilderness over the previous half-century, but nothing with the complexity of Mulholland's project had ever been tried.

Taylor's first task in mid-1908 was to make an inspection trip all the way up the line, one that was still about as difficult as it had been when Mulholland and Eaton had traveled it in 1904. There were still no paved roads north of Mojave, just a wagon track that was of little practical help to the newfangled automobiles that tried to make the trip. The first Model T had rolled off the assembly line in 1903, and by 1904, there were said to be 1,600 cars cruising the streets of Los Angeles, where the maximum speed limit was 8 mph in residential areas and 6 mph in business districts.

The Chief himself was not unfamiliar with such conveyances, for he had been a passenger in one of the early Oldsmobiles driven by George Read, chief of the water-meter division. According to Read, Mulholland had suggested they take a spin in Read's new contraption after work one evening and invited another department engineer, Fred Fischer, to ride along. As Read's narrative bears out, even at six miles an hour, the appearance of such a vehicle on the streets could create havoc.

"As I was driving around the Plaza," Read said, "a tall lanky fellow riding a bicycle approached us. I honked the horn, and not getting the fellow's attention, applied the brakes. I had just about stopped when he banged into the front of the car."

As Read described, the bicycle stopped, but its rider kept going, over the dashboard of the Oldsmobile, finally sprawling across the three of them jammed into the car's only seat. A shaken Read was concerned that the man might be hurt and that an intervention by police could be next.

Mulholland, however, was unfazed and fixed the man with a stare. "What the hell are you doing in here?" Mulholland demanded. In seconds, the man gathered himself, jumped back on his bicycle and rode away without a word.

Mulholland might have been adept with solving certain issues regarding urban auto travel, but there was little he could do about the obstacles to auto travel between Mojave and the Owens Valley. As Taylor pointed out, wagons were built with their wheels spaced sixty inches apart. Automobile axles, however, were only fifty-four inches wide. That meant that while one rubber tire could easily course along a rut formed by a wagon wheel, the other was traveling on top of an untended median. In muddy areas, the second auto tire was always bogging down, and in rocky terrain the ride became teeth shattering.

Even getting out of the San Fernando Valley was a chore in itself, as Taylor made clear. The unpaved San Fernando Road through the Newhall Pass rose at a twenty-six-degree grade and was full of stones and ruts carved by occasional cloudbursts. He could get up that hill only by periodically jumping out of the car to block the wheels from sliding backward, then gunning the engine into another lurch forward.

When they finally got over the pass and descended into Saugus, Taylor found that Mulholland had already seen to the erecting of a wooden hospital building at the project's southernmost division headquarters, near the Southern Pacific railway station. From there, Taylor and his men traveled up the San Francisquito Canyon Road, crossing the creek "about forty or fifty times" as they climbed.

Eventually, just south of the crest of the San Gabriels, they reached the camp at the South Portal of the Lake Elizabeth tunnel, still in much the same rugged condition that Chaffee had observed. An encounter with the belligerent foreman on the site suggested to Taylor that it wasn't of much interest whether medical facilities were established there or not, but in the end Taylor says, he "got a first-aid man in there who got along with him."

From the South Portal, it was a relatively easy drive over the crest of the San Gabriels and down across the Antelope Valley to Mojave, where he found a settlement similar to those depicted in contemporary Westerns. There was a Southern Pacific station on one side of the tracks and one of the familiar Harvey House hotels next to it. On the opposite side of the tracks were several saloons, a number of billiard parlors, two general stores, and a pharmacy, the latter run by what Taylor described as the only "medical man" in those parts at the time.

The pharmacy's proprietor had practiced medicine at one time, Taylor said, but "he was no surgeon." If anyone in Mojave were to be seriously injured, "they either died on the spot" or got sent on a train up to Bakersfield, about sixty miles to the northwest. The Harvey House was about the only place between Los Angeles and Independence where a decent meal was available, and it also featured clean beds, which at the time was no insignificant feature. There were rooming houses in Mojave, Taylor said, but all were vermin infested and none of his men ever stayed anywhere but the Harvey House if they could help it.

If Mojave was rough at the time of his first inspection, it only got rougher as the pace of work on the aqueduct grew and more and more men tumbled into the only settlement between Los Angeles and the Owens Valley. On payday, as any number of accounts testified, it was common for a digger to show up at a Mojave saloon with his paycheck, anxious for a drink and a good meal. It might be suggested at that point that there was not enough in the till to cash said paycheck, but the barkeep would be happy to keep it safe and simply deduct what had been consumed by the time the check could be cashed. Such an arrangement of course often led to the consumption of great quantities on the part of the digger, quite often leading to a state of unconsciousness sometimes assisted by the application of a blackjack or cudgel.

If a digger thus taken advantage of was fortunate enough to wake up—often in an alley somewhere—he was likely to learn that he had somehow managed to drink up the value of his entire paycheck, or that he had in fact forgotten that he had cashed it and must have subsequently been robbed. In time, Taylor would install a doctor at the headquarters camp just outside Mojave. This practitioner reported being called into town three or four nights a week to stitch up a fractured skull or pronounce someone dead. It was less than welcome duty for the attending physician, but as Taylor wryly pointed out, there was often a little something extra provided by a hotelier or saloonkeeper who just wanted help in "getting rid of the carcass."

About twenty miles north of Mojave lay the settlement of Cinco, where the headquarters of the rugged Jawbone Division would be located along the line of the new railroad. In order to maintain the proper level of descent toward Los Angeles, however, the aqueduct itself would be located in the steep hills several miles west of the rail line. Ultimately, Mulholland would have to build roads from the rail line up to the work camps, but, meantime, staging facilities had to be constructed at Cinco. Because of the complicated nature of the tunnel and siphon work required to move the aqueduct through this area, more men would work here than at any other place along the line, and eventually Cinco would operate the largest of Taylor's hospitals, twice the size of any other.

At another spring and former stage stop, Coyote Hills, Taylor met a settler named Freeman Raymond, a man in his sixties who said he had left the camp only twice in twenty-five years, once to get married. Coyote Hills lay at the foot of the last pass through the Sierra Nevada until one traversed the Owens Valley, and the camp was to become a favored stop for Taylor. There was enough water for Raymond to keep cows, and there were always eggs, ham, bacon, and chicken for Sunday dinner. With clean beds and cool water, Coyote Hills was a virtual oasis along the route.

From just above Coyote Hills, Taylor and his party were within view of the Sierra Nevada for the remainder of their survey trip, now past what would be the site of the big Sand Canyon Siphon, then up the gentle slope of the Rose Valley, and finally encountering what he described as nearly twenty miles of the worst sandy road along the entire route, "almost impossible to get through."

The only way to make progress at that juncture, Taylor said, "was to run your car as far as it would go, and when you got to a point where your wheels were beginning to slip . . . but before you killed your engine, you stopped, reversed, and backed up in the tracks you had made." Thereupon, one had to gun the engine and plow forward again, which might gain another stretch of fifty or seventy-five feet. Only in that way did the party finally reach Olancha, above the salty Owens Lake, which was about twenty miles long and ten miles wide at the time.

At nearby Cottonwood Creek, a sizable tributary of the Owens River, engineer Harvey Van Norman had already commenced work to construct a power plant to feed aqueduct construction. The work had been going well enough for Van Norman, though it was a continuing struggle to bring in supplies. The Southern Pacific's arrival in the valley was still a distant dream, and he had to rely upon the existing narrow gauge from the north that terminated at Lone Pine, sixteen miles away. The only means of carrying freight from that point was by mule team over a sandy road not much better than the one Taylor and his group had struggled over from the south.

The supply issue was one thing, but of much greater importance to Van Norman at the time was the question of what had happened to his wife. The two had only been married three weeks, Van Norman explained, and J. B. Lippincott had promised that the city would provide proper quarters for his new bride at Olancha. However, she had not yet arrived, and there had been no word from Lippincott as to her whereabouts.

Thus, Van Norman wanted Taylor to carry a message to Lippincott upon his return: if his wife hadn't shown up within a week, Van Norman would be down to hand in his resignation. "He was mad, and he would have done it," Taylor observed.

Luckily, however, Taylor and his men were to encounter the new Mrs. Van Norman making her way up the Nine-Mile Canyon during their return to Los Angeles, and, accordingly, the matter passed. Not only did Van Norman—a self-taught type in the mold of the Chief—continue on in the department's service, he would one day come to succeed Mulholland as chief engineer.

Before returning, however, Taylor and his party had traveled all the way up the Owens Valley to Division Creek, near the diversion point of the aqueduct. After a night's sleep atop a billiard table in an otherwise louse-infected ranch camp, Taylor made arrangements with the only physician in that part of the valley to care for the aqueduct men in Divisions One and Two, though the process of convincing "Doc Woodin" of Independence was not terribly difficult. Woodin was an old-timer who never gave anybody a bill, according to Taylor. Before his contract with the aqueduct, "If Woodin needed some money he'd just tackle somebody who owed him and tell him to dig up."

Woodin's idea of patient confidentiality was also less than delicate. "I've heard him holler at a farmer who had driven up on the other side of the street," Taylor recalled, "and ask him how his wife was and whether that last medicine he sent up did her periods any good." Another time, Taylor heard Woodin call after a new party of workmen as they were walking down the street, "Boys, Doc Taylor tells me to take care of you and send him the bills. But, don't forget, no clap—clap's barred." Given Woodin's capabilities, it was the sort of thing Taylor let go.

12

FIRST SPADE

TAYLOR'S COLORFUL ACCOUNTS UNDERSCORE NOT only the challenges associated with the project but the can-do attitude that came to characterize every phase of it. In short order, he hired physicians for hospitals at Saugus, Fairmont, and Cinco, and went to work staffing the lesser installations with what were called hospital stewards, most former army medics who had administered first aid in similarly primitive conditions. They could ride a horse, minister to colds and constipation, and even splint a fracture when they had to. In all, there would be six hospitals with physicians attending and sixteen first-aid stations manned by stewards, along with sixteen horses and two motorcycles at Taylor's disposal. If any truly serious case cropped up, a sick or injured workman could be put aboard a train and carried to California Hospital in Los Angeles for treatment, a process that became increasingly efficient as the Southern Pacific Line pushed northward out of Mojave.

Meantime, with medical matters attended to, Mulholland was anxious to begin work in the Jawbone Division, where the task would be as challenging and time consuming as tunnel work at Lake Elizabeth. However, until the bonds could be sold, there was no money to pay for workmen or supplies.

Though the City Commission had authorized city attorney Mathews's suggested increase in the bond rates from 4 to 4½ percent, San Francisco had an issue of $18 million at 5 percent still begging for takers, and New York City had upped its rate on a $20 million issue to 5 percent as well. Finally, an offer came to the city from one local firm willing to purchase $2 million of the bonds, followed quickly by an identical offer from another local, N. W. Harris, purchaser of the initial bond offering of 1905. While the city pondered these stop-gap offers, city attorney Mathews returned from New York with the welcome word that a syndicate headed by Kountze Brothers and A. B. Leach & Company had offered to buy $4 million of the bonds outright and take up an option on the remainder.

As the *Los Angeles Times* reported, the City Commission voted its approval of the sale on July 10, 1908, the largest such offering in its history. Chaffee told reporters that it was "a lucky sale, all things considered," adding, "Now we are going to show you how to make the dirt and rocks fly." For his part, Mulholland pledged, "the waters of the Owens River will be flowing into the San Fernando Valley by July, 1912."

Though Mulholland had already agreed to let bids out for the work in the twenty-two-mile long Jawbone Division, the responses from private companies had been as dismal as he predicted. One private contractor who surveyed the area prefaced his submission by saying that he had never in his career seen conditions so adverse to such an undertaking. Even allowing for a 15 percent markup to contractors off the top, Mulholland said, the estimates that came in were beyond reason. As a result, he said, the city would be doing its own work there, including twelve miles of tunnels and eight siphons, and for about half the cost of the lowest bid that had been proposed. On average, Mulholland figured that the cost of each of the 1 million or so feet in the entire system would average about $22. By comparison, figures from those who bid on the Jawbone work came in at more than $50 a foot, a figure that would have bankrupted the project before it began.

By early August, shortly after the last of the Jawbone bids had been rejected, Mulholland had 400 men at work in the area; by fall there were 700 there, and by the beginning of 1909, almost 1,300. As Mulholland told reporters, "Any man of family in Los Angeles who desires work at manual labor for $2 a day can secure a place," adding that there would be "modest" homes built—some already available at Tehachapi near the cement plant—as well as schools, one already proposed near the South Portal—so that wives and children of workers could be brought along. Though such facilities added cost to the project, Mulholland argued to the commissioners that family men were more reliable than the itinerants who were content to live in tents.

Despite such blandishments, however, the number of men who brought their families to live in aqueduct camps remained few. One later study calculated that only twenty-three women and twenty-three children ever lived in the Inyo County camps. The women who did come were wives of skilled or supervisory workers, although two prostitutes were also enumerated.

Part of the knottiest construction problem in the Jawbone, where he would eventually have fifteen separate crews working, Mulholland told reporters, was the amount of preparatory work involved. Several miles of roads leading to the route of the aqueduct from the mail trail between Mojave and the Owens Valley had to be carved, telephone and power lines strung all the way down to the isolated region from the Cottonwood Creek power plant in the Owens Valley, and water pipelines twenty miles long laid from wells near Tehachapi and in the nearby mountains. Meanwhile, however, Mulholland announced that a mile of canal had been dredged in the Owens Valley and was brimful of river water; the power plant at Division Creek was at work and feeding electricity as far south as the Alabama Hills; and nearly a mile of the Elizabeth Tunnel had been bored. There was no question in his mind that the big tunnel through the "nice" rock down there would be finished by the time the rest of the line was in place.

One of the innovative features of the nascent project was the floating dredge eating its way along near the Alabama Hills. There had been somewhat similar dredges used before, but "Big Bill," as workmen named this one, had been designed by Mulholland and, using powerful streams of water to cut through the mud and soft earth in its path, was said to be the most efficient of its kind. "Mr. Mulholland thinks as much of that dredger as though it were his baby," one of the stenographers in the aqueduct offices told a reporter.

All the while, there was an existing water system in Los Angeles to be maintained, one that was constantly increasing in its demands. When the city originally took over the works, Mulholland pointed out, there had been only about 25,000 customers. By November 1908, there were nearly 60,000, with about 300 new accounts being added each month. When he began as city superintendent, there were about 650 meters in use, a figure that had grown to more than 25,000. In the coming year, he hoped to more than double that number. Despite the savings brought about by the metering system, the city's consumption of water had risen from

23 million gallons per day to almost 45 million, a reminder of why the completion of the aqueduct was vital

At the beginning of December, Mulholland submitted his "Third Annual Report" to the Board of Public Works, summarizing the work completed to date and projecting that fully fifty-eight miles of the line would be finished in 1909. He lamented the time and expense involved in all the preliminary work as admittedly "appalling," but stressed that "it was felt that better and more economic progress could be made with all the accessories in the completed state in which they now are."

He assured the board that morale was high among the workforce and supervisors alike, and that there was every reason to believe that the project would be completed "within the time and cost originally estimated." Within the crucial Jawbone Division, he noted that about a mile and a half of tunneling work had been completed and that most of the rest would be finished in the year to come. Progress had been steady on the tunnel beneath Elizabeth Lake, he added, with the North and South Crews moving toward each other at the rate of 660 feet per month.

At a meeting of the Chamber of Commerce later that December, the chairman of the group worked up the temerity to ask if Mulholland could give them some better sense of just how much work had been done to date on the aqueduct. Mulholland paused, glancing about the room with a look that suggested he'd finally been uncovered as a fraud.

"Well, we have spent about $8 million all told, I guess," the Chief answered, "and there is perhaps nine hundred feet of aqueduct built. Figuring all our expenses, it has cost us about $3,300 per foot."

Coming from the man who had estimated that he could build the entire project for an average of $22 a foot, the statement would have sent murmurs through the room. A third of the money gone and only 900 feet of aqueduct to show for it? Mulholland simply nodded, giving the audience plenty of time to let the figures sink in.

Then, with his characteristic sense of timing, he brightened. "But by this time next year," he said, "I'll have fifty miles completed and at a cost of under $30 per foot, if you'll just let me alone."

With that, the audience dissolved into hoots and applause. "All right, Bill," the chairman laughed. "Go ahead; we're not mad about it."

And go ahead Mulholland did, given that there was finally some money in the kitty. He laid out plans for work that would total nearly $5 million, reorganizing activities into thirteen divisions, including the splitting of the Antelope Valley work in two and setting aside work at the cement plant as a division unto itself. For the coming year he planned significant work in ten of those divisions.

He had refined the workings of "Big Bill," the dredge operating between the intake and the Alabama Hills, and intended to keep it running night and day during the upcoming summer months. They would use a land-based power shovel to dig through the more rugged terrain at the Alabama Hills, with construction gangs to follow along, shaping the open line there by hand. With the onset of warmer weather in April, another power shovel would go to work excavating the reservoir site at Haiwee and other crews would begin digging northward from Haiwee toward those working in the Alabama Hills.

The work in the Rose Valley, directly below the Haiwee Dam, consisted of fifteen miles of relatively easy work and thus could wait. The next section, though, which Mulholland now dubbed the Grapevine, was "rough and expensive," and would require a number of steel siphons and a significant amount of tunneling through solid rock. The Southern Pacific rail line from Mojave had finally reached the Grapevine area, however, and he planned to put two power shovels at work there during the summer once a dependable water supply was established.

The Freeman Division, south of Grapevine, was a twenty-one-mile stretch of what he called "cut and cover work of easy character." Nor did he anticipate any difficulties in getting the line extended across the Antelope Valley. The southern half of that stretch had just been let out by the city to subcontractor Perry Howard, who promised completion within two years. Mulholland already had two steam shovels on his own twenty-eight-mile section of the work in the northern Antelope Valley. In the Mojave Division, as he called it, he had originally intended to run an uncovered conduit similar to that between the diversion point and Haiwee, in the Owens Valley, but the work had gone so smoothly so far that with the savings already realized he could afford to cover that Antelope Valley portion "so no jackrabbits fall into it," and still come in under budget.

In the Elizabeth Division down south, work continued at both ends of the great tunnel, and excavation for the Fairmont Reservoir was under way. Mulholland also hoped to begin work on the Fairmont Dam itself, but his engineers had not yet settled the question of whether the structure would need to be built of earth or concrete. The work below the South Portal of the Elizabeth Tunnel was now known as the Saugus Division, with the charge of getting the aqueduct down San Francisquito Canyon and through the major tunnel at Newhall. It was "very rough" going there, Mulholland told a *Times* reporter, and work on twelve lesser tunnels would begin shortly, though any of the easier surface digging could wait.

The most challenging part of his plan, however, lay near the aqueduct's midpoint, within the twenty-two-mile-long bounds of the Jawbone Division, where crews had been busy at preparatory work since September 1908. Mulholland had budgeted more than $1 million for work in the Jawbone alone, planning to put an electrically powered shovel to work there and to complete most of the 43,000 feet of tunneling work before the end of the year.

It was at Jawbone where the *Out West* reporter early in 1909 climbed to a vantage point 1,200 feet above the desert floor to stare down upon "a yellow streak of excavated rock and dirt and the thousand ants of men at work, the line appearing and disappearing in a country torn and twisted and tossed in the eruptive period of ages gone," there to wonder upon Mulholland's "effrontery" in setting the project in motion to begin with.

By that time, the Southern Pacific had forged its line two-thirds of the way up from Mojave to the Owens Valley; 200 miles of access roads had been built; two power plants were up and running, with lines connected to the bulk of the construction sites; a telephone line stretched from the Los Angeles headquarters all the way to the intake site; the cement plant was churning out a thousand barrels a day; and machine shops, barracks, commissary buildings, warehouses, offices, and hospitals now dotted 230 miles of once-barren landscape.

In the first ten days of 1909, Mulholland's men drove a record 2,456 feet of tunnel in the Jawbone Division, and by February he calculated that progress up and down the line had reached the rate of 760 feet a day, or about 4⅓ miles per month. With new machinery coming on line, Mulholland hoped to reach a rate of five miles a month by summer.

One newfangled contraption that Mulholland put to work on the project would become legendary: a steam-powered traction engine developed by the Benjamin Holt Company, one that had enjoyed some success in helping develop the treacherous peat bogs near Stockton. The theretofore little-known device resembled a cross between a locomotive and an army tank with continuous steel treads placed where ordinary wheels would have been. Given the distances involved, the difficult terrain, the extremes of heat and cold, and the weight and volume of steel pipe and other supplies to be carried, Mulholland eventually agreed to commission the building of twenty-eight of the machines for service up and down the line to supplement the traditional mule teams that had been used for such purposes.

Water department lore has it that Mulholland watched one of the devices at work shortly after its delivery at the South Portal in 1907 and remarked, "It crawls like a caterpillar," thus christening the machine. That it was in fact the Chief who coined the phrase is unlikely (contemporary company literature attributes the name to a Holt photographer), but these "traction engines" showed great promise when they were first put to work and were just one more aspect of a project that seemed otherworldly for most. Widely featured in news accounts and photographs of the undertaking, the "caterpillar" would in time become a standard fixture on any earth-moving project.

The *Out West* writer also spent a fair amount of time watching Mulholland at work in the field. The Chief might observe a gang at work with pick and shovel for half an hour without saying a word, the reporter noted, then later take the foreman aside to offer advice on how to improve his gang's efficiency, along with a deft assessment of who was worth keeping and who was a drag on the line.

It was the rare instance where Mulholland ever second-guessed a foreman, however, for he figured that a man in charge would never have achieved a position of authority if he hadn't merited it. On one occasion, Mulholland sent a laborer out to a job with a personal note of recommendation. A short while later, the reporter noted, the laborer was back in Mulholland's office.

"What are you doing here?" Mulholland asked.

The laborer threw up his hands. He'd presented Mulholland's

letter of recommendation, but the foreman had told him he had all the men he needed and that all of them were doing good work besides.

Mulholland nodded. "I guess that settles it, then. Big John is responsible for laying that pipe, and if he says there isn't a place open, then there isn't."

Finally, the reporter who had wondered what had made Mulholland think he could pull off such a massive undertaking confronted the man himself with the question, one that drew a rare concession. "I don't know why I ever went into this job," Mulholland answered with a wry smile. He admitted that there was far more money to be made in a private practice, but then he paused. "I guess it was the Irish in me."

And there was also the fact of his ingrained sense of duty, he said. "I know the necessity, better perhaps than any other man. . . . If I don't, my thirty years of employment on the city's waterworks haven't gone for much."

Mulholland sat silent for a moment, then gave his questioner a final reassuring nod. "We'll pull her through on time, never fear, if the men in the ditch can have their swing."

BEST YEAR TO DATE

THE LIST OF PROBLEMS WAS NEVER SHORT FOR MULHOLLAND. If the work itself were not enough, he had long been irked by the general intransigence of the city's Civil Service Commission, a matter that came to a head when that body insisted that J. B. Lippincott, Mulholland's chief assistant on the project, be required to pass the appropriate engineering Civil Service Examination or step down. To a practical-minded man such as Mulholland, the edict was the height of foolishness.

"There is always the danger that he will not stand first in such an examination," Mulholland testified in a hearing before the commission. But Lippincott, the Chief said, "knows what he is doing from the ground up," in contrast to even top engineers, "who though they might best him in an examination, would be vastly inferior in knowledge of the work at hand."

Adna Chaffee appeared at the hearing along with Mulholland and also argued for an exemption for Lippincott. In the end, the commission requested that Mulholland and Chaffee provide a revised classification and definition of aqueduct-related duties. Meantime, they said, they would take the Lippincott matter under advisement. It should be noted here that if the council and new mayor thought the appointment of Adna Chaffee would create some sort of check on William Mulholland, they would be sadly mistaken. The fact was that throughout the arduous days to come, these two warhorses proved to be kindred spirits, united in their disdain for petty politics and motivated primarily to get things done.

There was another brief flurry of concern in February when the California legislature considered a bill that would increase the minimum wage to $2.50 per day on all public works projects in the state. Given that the main budget item in his estimates was for labor, Mulholland fretted that if the measure passed, it would mean a $3.5 million increase in payroll alone. Ultimately, however, and owing to the lingering depression and high rate of unemployment, the measure was defeated in Sacramento.

At the Jawbone, meantime, engineer John Freeman remarked to his chief on a survey trip that the place appeared to be nearly impossible country for canal digging. "It *is* rough on top," Mulholland agreed, adding that it was the reason they were going to do much of their work there underground. "When you buy a piece of pork," he pointed out, "you don't have to eat the bristles."

Tunneling carried its own version of "bristles," of course. The Coastal Mountains of California are traversed by multiple fault lines, and underground digging and blasting in such terrain, even where the work seems to be going on in solid rock, can often trigger unexpected cave-ins or loose a flood of suffocating sand from hidden pockets. In those days, there were no metal hard hats, but miners did wear derbies that cushioned the blow of a falling rock. As one worker explained, "Any tool dropped from a height not too great, or a falling rock not too large, would cave in the derby or drive it down over a man's ears, but not crack his skull."

To speed the pace of the treacherous tunnel work, Mulholland instituted a bonus system. For every foot that a crew exceeded the posted daily average—six feet where the roof had to be shored up with timbers, eight feet where it didn't—each man would get forty cents added to his day's pay. It was only one more reason why men were eager for what otherwise might have seemed dangerous and difficult for most.

Fredrick Cross, a nineteen-year-old surveyor's assistant who helped chart the course for the tunnel men, recalled that he accepted underground work with "the hazards of falling rock and delayed charges of exploding dynamite, the annoyance of the deafening air drills and the continually dripping water," primarily because it got him off the alternately blistering and freezing desert floor. In the tunnels, the temperature held at a steady fifty-eight degrees, Cross said, "with no burning sun in the summer, no biting cold in the winter and no gritty winds."

Cross was eloquent in his recollections of his time on the job at the Jawbone. He alighted from the Southern Pacific's *Owl* in Mojave on a July night in 1908, stepping out onto the platform in the midst of a gale whipping down from the Tehachapi Pass, paper alligator-hide suitcase in hand. Hardly had the brakemen called out to the departing passengers, "Hold on to your hats!" then Cross felt his brand-new five-dollar Stetson lift off his head and go wheeling into the moonless desert.

He made his way to the nearby Harvey House where the barmaid gave him a knowing look as he slid onto a stool. "So you donated a hat to the Hat Ranch," she said, wiping the bar before him. "You know, that's how the Indians around here get their hats."

If it was the first jibe he would have to endure as a tenderfoot, it would scarcely be the last. "Being able to adjust an instrument or throw a chain into a perfect circle," as Cross would soon learn, "did not guarantee acceptance" by the old-timers on the job. The surveying crew that he joined up with at Cinco contained a motley collection of sun-bronzed veterans and immigrants—including a significant number of Greeks, Bulgarians, Serbs, Montenegrins, Swiss, and Mexicans—who had worked railroading and mining jobs throughout the West from the days when army troops had to be stationed to prevent Indian attacks. One earned one's spurs among such men by lugging heavy surveying gear up a desert hillside without constant canteen breaks, being willing to dangle from a rope off the side of a hundred-foot cliff with seeming indifference, and maintaining a passive face at the poker table, even with aces in the hole.

For his work, Cross drew a monthly paycheck of seventy dollars, with thirty of it deducted for food. He sent another thirty dollars home to his mother, which left, as he put it, "ten to spend on shoes and riotous living." But the food was good, he said, at least in the surveying camps. At the larger mess halls along the line run by Joe Desmond's company, however, the fare tended more toward the dismal. Starches formed a disproportionate share of the food triangle there, and the meat was always tough. A favorite complaint of the tunnel men was that at Desmond's mess, "the pies have a pay streak that's too thin."

While Cross was a slight grade above the bottom in classification, it was nonetheless the "stakemen," or laborers, who formed the backbone of the effort up and down the line. They loaded or "mucked" rock onto cars, then pushed them out of the tunnels, cleared the surface route of creosote bush and cactus, and followed like drones in the wake of steam shovels, using picks and pry bars and hand shovels to shape the ditches. By the end of March there were nearly 3,000 men at work on the aqueduct, most of them working hard for their $2 per day.

Project statistics show that the average laborer did not stay on such a demanding job for more than two weeks at a time, however. Since bonuses were paid out in ten-day increments, that payout was usually the occasion for a significant number of men to simply walk off the job, headed either for Mojave or for one of the so-called rag camps, or tent saloons, erected by entrepreneurs in the intervening stretches of desert. It was a standing joke among foremen on the line to regularly report having "one crew drunk, one crew sobering up, and one crew working."

Mulholland once allowed that it was the rare laborer who would endure the conditions for long "if whiskey didn't keep him broke," an observation that young surveyor Cross seconded. He had often seen a man take his pay and strike off southward, vowing that he was done forever with work on the blasted aqueduct. In a week or so, the man would return, begging for his job back, having made it only as far as Mojave or the 18-Mile House, a saloon erected hastily in the desert near Cinco. As Cross put it, such a "hard rock" was not ordinarily a gambler as well. "He just wanted whisky. It made him forget for awhile."

Cinco, the staging area for work in the Jawbone Division, sprang up along the rail line where there had been nothing but jackrabbits and creosote bush before. Erwin Widney, who signed on the project as a timekeeper shortly after his high school graduation, pointed out that something might indeed come from nothing, for Cinco was now "a great concentration camp for men and materials, both for the aqueduct and railroad." When he stepped down from his car the morning of his arrival on the job, he saw nothing for hundreds of yards in every direction but stacks of ties, rails, crated spikes, spurs lined with tank cars, flat cars, and ore cars piled high with rock and gravel.

Out into the desert stretched an assemblage of quarters-tents, larger mess tents, rough-boarded shops and storage buildings. Men were busy loading mule-drawn freighter wagons with gallon cans of black powder, crates full of dynamite, feed, kerosene, picks and shovels, and all manner of supplies for satellite camps and work stations arrayed in the distant hills, where the flat line of the distant aqueduct was just visible from where he stood.

Though the work would prove rugged and beyond taxing, there were also the elements that form part of the enduring intrigue and exoticism of the West. For all the hardships, surveyor Cross spoke of the beautiful side of life on the Jawbone: "A windless night seemed to bring the stars closer; the full moon gave light enough to read newsprint" (newspapers were generally two to three days old, arriving on the same thrice-weekly schedule as the mail). No matter how hot the day had been, Cross added, sundown in the dry desert always brought with it relief.

"There were quail, dove, cactus wren and raven," Cross remembers, "jack rabbit, tortoise, coyote, chuckawalla (lizards), and the ubiquitous rattlesnake—both diamondback and sidewinder." On one occasion, Cross and his companions came across a huge beehive nestled into a crevice in an otherwise barren cliff side, providing them with a cache of sweet wild honey the likes of which they hadn't had for months.

Widney recalls the vivid sight of the first mule skinner he encountered on the line, a man in his fifties, handsome "in a renegade and dissipated way . . . a man anyone would turn to look at a second time." The man seemed called out of Central Casting in a day before Central Casting existed, dressed in the height of skinner fashion, including a large Stetson with a silver-studded leather band, a blue denim shirt, and bright scarf, and "a beautiful silver-studded leather vest and cuffs to match, the latter reaching from the wrist nearly to the elbow."

In new blue denim pants and napa leather boots, the skinner stood watching a pair of helpers carry out the loading, occasionally calling out a bit of advice. This was a man too well dressed to be a "common" mule driver, Widney thought, noticing that he also seemed to be "pleasantly under the influence of liquor." In time, Widney would learn that the driver he saw that first morning on the Cinco platform was indeed a legend among his peers. Though most of the skinners dressed modestly, did their own loading, and stayed sober, at least during working hours, this man could handle a team of fourteen mules with ease and was unfazed by running his wagons over treacherous mountainside roads to the highest camps, including certain routes that no one else was willing to brave. "He does it drunk better than sober," one of the other skinners told Widney, and thus he was left alone, "to do very much as he pleased."

Because of its location, Cinco was something of an anomaly. Much more typical of the work camps was that known as 30-A, a few miles south of Little Lake, where Widney was permanently assigned. That camp consisted of a half-dozen boxcars that had been converted into a combination bunkhouse/kitchen/dining room/storehouse/commissary, one corral for mules, and an adjacent tent that was the blacksmith shop. The "bunkhouse" was frigid in winter and an oven in the summer, though on mild nights, many of the men took their cots up to the running boards atop the cars. (Men provided their own cots, pads, and blankets.) "There is no place as fine as a high desert plateau for surveying the celestial kingdom," Widney recalls, "and lying on your back on a comfortable cot with no enclosure but the great limitless universe is the most pleasing manner to enjoy it."

Despite the occasional diversions, the divisions where Cross and Widney were assigned were doing the most difficult and dangerous work. Inside the tunnels themselves, three shifts worked steadily at their tasks in the light thrown by candles made of stearic acid. There would be nine holes drilled with the proper spacing, a stick of dynamite inserted in each, fuses run, then lighted.

"Fire" was the call as everyone ran to the portal to wait. When the explosions came, the men counted the number carefully. As Cross put it, "No man wanted to drop a pick into a live stick of dynamite." The moment the air cleared, the ore cars were run back in, and the muckers began shoveling. It was a laborious process, one that repeated itself endlessly. While the holes for charges were initially bored by hand, and muckers themselves pushed the rock-laden cars up the exit tracks, the arrival of electricity in the various locations speeded up the process considerably. For the men themselves, power drills and motorized cars transformed the basically intolerable to the merely exhausting.

As challenging as the tunneling was, so was the work to build the massive steel siphons used to carry the water across certain of the deep canyons, including the most formidable at Jawbone Canyon itself. (There would be twenty-three siphons constructed in all, eight built of reinforced concrete, the rest—except for one at Sand Canyon—of steel plate, ranging in thickness from 1/4 inch to 1 1/8 inches.) First, the siphons had to be carefully designed in accordance with length and distance and rate of drop and rise, and even then there was a certain amount of dead reckoning going on, as Mulholland admitted. He thought these structures would work, but nothing of such scale had been done before.

The Jawbone Siphon itself was drawn up to be a little more than 8,000 feet long and 10 feet in diameter in places, and to operate under a pressure of 365 pounds per square inch (water pressure in a typical modern home comes out of the faucet at 50 pounds per square inch). In some places, the walls of the canyon slope upward at a grade of 35 degrees (by contrast, the maximum grade allowable on the US Interstate Highway system is 6 percent). That all might sound daunting enough to the nonhydraulic engineer, but to Mulholland, the principal issue was to come up with a siphon light enough to be held in place as it traversed such terrain yet still strong enough not to rupture under pressure.

"Theoretically," he said, "the most economical pipe . . . would be a tapering one, the large diameter at the top, where there is no pressure, and the small diameter at the bottom." While it was impossible to construct such tapering pipes at the time, Mulholland got around the problem with a design that stepped down the diameter of the pipe at several places where the Jawbone Siphon descended the canyon wall. As the pipe went up the other side, it would increase in diameter correspondingly.

Once all the calculations were made, drawings were sent to foundries in Pennsylvania and New Jersey for fabrication. Sections of pipe would be freighted by flatcar to Los Angeles in minimum lots of 30,000 pounds, then sent on up the new aqueduct supply line to be stored at the appropriate siding.

It was at that point that matters became tricky: How to transport a section of steel pipe 37 feet long and 1⅛ inches thick, weighing 52,000 pounds, up several miles of dirt road? In that particular case, it was a section of pipe to be used in the Jawbone Siphon, its size outstripping even the abilities of the caterpillar engines. Mulholland and Harvey Van Norman finally managed it the old-fashioned way, devising a pair of huge wagons with tires two feet wide pulled by a team of fifty-two mules.

Once at the work site, however, the difficulties continued. How, for instance, to lift a twenty-six-ton section of pipe up the side of

Los Angeles' second waterwheel, lifting the city's supply from the Los Angeles River to the old Sainsevain Reservoir during the Civil War era.

Downtown Los Angeles (at the northwest corner of Second Street and Broadway) in 1890, the year of William Mulholland's marriage to Lillie Ferguson.

The headwaters of the Owens River as Mulholland would have discovered it, carrying 26 million gallons of water or more each day.

Cottonwood Creek work camp in the Owens Valley with snowcaps in the background.

Lining a section of the aqueduct in the Owens Valley.

Mulholland's favorites: "Hayburners" in the harness hauling one of the massive sections of the siphon pipe.

An early "Caterpillar" on the job.

Keeping to the grade: A tunnel crew atop their muck cars.

A portion of the lined aqueduct below the aqueduct intake in the Owens Valley.

Power shovel love.

The jaw-dropping Jawbone Siphon in 1913.

Workmen at the ever-dangerous job of coating the interior of aqueduct pipe.

Maybe Maude.

Out of the future: The interior of Power Plant #1, San Francisquito Canyon.

The completed aqueduct just above a still-full Owens Lake.

A river tamed: Opening the gates at the diversion point 12 miles north of Independence, California.

"There it is. Take it": Opening the gates at the Cascade below the Newhall Pass, November 5, 1913.

Mulholland (*in the foreground*) with water commissioner Reginaldo del Valle in a section of 80-inch pipe during the construction of Power Plant #1 in San Francisquito Canyon, circa 1916.

Sabotaged aqueduct pipe disgorging water into its former course, at No Name Canyon, May 27, 1927.

Grandeur: The St. Francis Dam as it appeared shortly before its collapse in 1928.

And then dismay: The remains of the St. Francis Dam shortly after its collapse, March 13, 1928.

Dazed Santa Clarita Valley residents comb the wreckage downstream from the St. Francis Dam collapse.

Power Plant #2 as it appeared in 1928, shortly before the collapse of the St. Francis Dam.

The remains of Power Plant #2 following the St. Francis Dam collapse.

ɪulholland (*left*) with
ɪeorge Bejar at the site of the
t. Francis Dam shortly after
s collapse, March 12, 1928.

ɪely at the top: Mulholland in
923, at sixty-eight, backpacks
ɪ the wilderness to survey the
ute of the Colorado Aqueduct
from Boulder Dam.

What has come:
The Los Angeles River today.

Not Des Moines: The iconic night view of the San Fernando Valley from Mulholland Drive.

Legacy: Mulholland Dam and the Hollywood Reservoir as viewed today from the nearby Hollywood Hills.

The penstocks of Los Angeles Aqueducts 1 and 2 at Newhall Pass.

a thousand-foot cliff, maneuver it into near vertical position, then rivet it to the previously installed section? At Jawbone Canyon and others of the most difficult locations, incline railroads were built to haul the pipes up to otherwise unreachable spots, and aerial platform and gondola lines were erected in order to carry workmen between cliff top and desert floor. A system of winches and heavy rope was used to lift and hold dangling pipe sections while they were being riveted into place.

Even the riveting process itself was laborious. While the rivet holes had been punched at the foundry and carefully matched up prior to shipment, each rivet could be as much as an inch and a half in diameter and weigh up to five pounds. Each had to be set by hand and driven in place by an electrically powered compressor hammer, a process difficult enough when pipe was lying horizontally on solid ground, let alone dangling on the side of a cliff. Along with the development of a workable tunnel-drilling routine, an efficient riveting process was crucial to bringing the project in on time and within budget. Accordingly, Mulholland instituted the bonus system for riveting crews as well, and the ever cost-conscious Chief was happy to report that crews working under the bonus schedule were able to increase their average process from 7,725 rivets a month to 11,459.

Joining the sections of siphon and pipe was arduous, but keeping them in place, even for the horizontal sections of pipe, proved to be another problem. On the desert floor, where the grade was reasonable, Mulholland first tried burying the pipe in trenches, but the cost of excavation—as much as $10 a foot—proved to be prohibitive. There also proved to be another issue: given the high mineral content of the soil and the inevitable leakage at joints, buried pipe rusted more quickly and, of course, could not be easily spotted, repaired, or recoated.

In the end, Mulholland decided that nearly all of the steel pipe used along the line would have to be laid above ground, on supporting piers, with a typical span averaging about twenty-four feet, an approach that in the end proved far cheaper than trench digging. But the pier system also involved a bit of trial and error. Initially, it was thought necessary to extend the supports in a semicircular fashion, cupping the pipe to a point about halfway up the sides—almost like a series of concrete, metal, and wooden hands cupping the pipe as it crossed the brutal terrain. The first such system was constructed at the Nine-Mile Canyon Siphon in the Grapevine Division to the north.

However, when a section of the pipe there was filled in order to test the strength of the arrangement, there came an unexpected result. The supports proved sufficiently sturdy to hold up the load, but the pressure of the water on the sides of the pipe was so great that it flattened and bulged outward, distorting what had been a circle of steel plate into an oval. The tops of the cupping supports were sheared off in an instant, catapulting pieces into the surrounding desert.

Even at age fifty-four, Mulholland realized, there were always things to be learned. Following a survey of the wreckage, he said, the supports would be built only as high as what the grade called for. Once the pipe had been assembled atop the supports and filled with water, and only then, would the support sides be added, thus acceding to the final shape that the pipe itself had taken. Man would have to adapt to the will of the iron pipe, not the other way around.

Even with such setbacks, Mulholland was happy with the way the steel piping process shaped up. He would be paying the mills a little less than two cents per pound for the pipe, and even with transcontinental freight, plus the freight bill for hauling up to Cinco by the Southern Pacific, plus the cost of mules, wagons, and hay, and with engineering, equipment, and general field overhead also figured in, he calculated that he was laying pipe over some of the most god-awful country in California for a total of four cents per pound. In contrast, commissioners had let a private contract for two small siphons in the Saugus Division for five cents a pound. Just more proof in Mulholland's eyes that in aqueduct matters, they would do better to listen to him.

While the difficult and tedious laying of pipeline continued in the Jawbone and elsewhere, some of the most spectacular work in 1909 was within the tunnels all along the route. In August 1909, crews set a world record for soft-rock tunnel driving in Tunnel 17 at Jawbone, cutting through 1,061 feet of sandstone in a month. Earlier, a crew at the Elizabeth Tunnel broke an American record for hard-rock tunneling by boring through 604 feet of granite in a month.

Such accomplishments, Mulholland was quick to remind anyone who would listen, were attributable to the effects of the bonus system. When he was hauled before the city auditor to explain why the November bonus payrolls were so large, Mulholland patiently explained. His men had dug a total of 9,131 feet of tunnels during the period, almost twice the normal amount. The total wages paid amounted to $9.87 per foot. If there had been no bonus system in place and had the men dug only what they were required to by contract, the city would have paid out $13.80 per foot. Not only had the bonus system saved about $4 a foot in costs, it was driving construction forward at record rates. The groundbreaking system not only underscored Mulholland's ingenuity, it also exhibited his ability to adapt and to maintain the loyalty of a huge workforce despite grueling conditions. He was not slave-driving his men, but rather was making them masters of their own destiny, an approach that would mitigate against labor problems for the life of the project.

Much of the encouraging news had been passed along to the

citizens of Los Angeles by newspapers, including a two-page splash in the September 9, 1909, *Los Angeles Times* by Allen Kelly, who had been covering the project for the paper from the outset. Kelly, who worked as an engineer's assistant early in his career, accompanied Mulholland on a trip along the route before construction began in 1907. At that time, he said, there was nothing more than a camp or two in the mountains, and he repeated a common theme when he wrote that it was "a notable feat of the imagination to call up anything like a credible mental picture of a great stream of water flowing across those vast stretches of desolate plain and under leagues of jumbled mountains."

By the fall of 1909, however, all had been transformed, Kelly said: "That vast region of solitude and desolation has been converted into a humming hive of human industry." He doubted that he could do more than hit the highlights, Kelly said, but he gave it his best. Though nearly all was preparatory work in the year between approval of the $23 million bond issue and October 1908, Kelly reported that in the eleven months since, there had been twenty-two miles of tunnel dug, sixteen miles of concrete-lined canal finished, and four miles of headwaters ditch dug in the Owens Valley. As a result, he said, there would be Owens Valley water flowing to the San Fernando Valley by 1912, a year ahead of schedule.

Kelly went on to detail how $8 million had been expended to date, about half on the preparatory work and much of the rest spent over the previous year. One of his more poetic descriptions was of the Chief's pet dredges at work on the big ditch in the Owens Valley, making them sound like creatures out of H. G Wells: "They float in the canal, cave down the ground with hydraulic nozzles, discharge the mud on both sides, and walk along on spud legs, towing transformers behind them."

Kelly noted that with the completion of the Division Creek

and Cottonwood Power Plants under engineer Scattergood's supervision, the city was now paying for project power only on the Saugus and Elizabeth Divisions. And he shared Mulholland's observation regarding the unexpected ease with which the tunneling through the Jawbone Division had been proceeding: at the rate things were going there, the Chief had told him, and if it had been sandstone underground all the way between the Owens Valley and Los Angeles, the entire line of the aqueduct could have been made of one great tunnel, and for less money than was budgeted in the bargain.

In the difficult Saugus Division, two and a half miles of tunnel work had been completed, and a shaded alpine settlement of several hundred had grown up near the South Portal of the Elizabeth project, where electric power was now being used to haul the blasted rock out of the bore. All twenty-eight of the caterpillar traction engines were now in place along the route, though 650 mules also shared the work of hauling.

All the progress was due to "the creative genius and engineering skill of Mulholland," said Kelly, adding that Mulholland's method of supervision also seemed remarkable. "I have seen him sketch with a stick in the sand the outline of a piece of work or a mechanical device for the man selected to do the work and then leave the man to work the thing out in detail." According to Kelly, Mulholland's elaborations were simple: "There is the principle. Apply it in your own way."

Kelly claimed never to have seen the superintendent with a notebook in his hand. He had never seen him make a written note or consult a document when asked to provide a statistic. Yet he could break out almost any cost detail of the project for the Board of Public Works: of the untold thousands of rivets to be driven into pipe along the route, Mulholland said, small ones could be set for five and two-thirds cents apiece and large ones for twenty-seven cents. Likewise, he could translate the cost of a bale of hay into the number of feet that a mule who had eaten that hay was able to pull a pipe-laden wagon. The instruments he used to survey and plot the original 235-mile route of the aqueduct consisted of a pocket compass and an aneroid barometer. If he'd been born in another time or place, the reporter theorized, Mulholland might have been a general and achieved "prodigies of destruction instead of construction."

With the work finally well underway, J. B. Lippincott arranged a kind of summit meeting of the various division supervisors to facilitate the exchange of ideas, and as could have been expected, the assistant arranged for his Chief to deliver the keynote. Those present included W. C. Aston, the grumpy foreman who was less than happy to welcome Dr. Taylor into his Elizabeth Tunnel camp, Harvey Van Norman, in charge of the Owens Valley Division, and Tom Flanigan, whose crews had set the soft-rock drilling record in the Red Rock Division. Mulholland's charge was to speak to his managers on "The Organization and What Is to Be Expected of It."

Mulholland rose, glanced about the room, and began, "I have got what I expected of the organization: loyalty and efficient work." He nodded for emphasis, then continued. "When I have been impatient and have criticized you in my rude and rough way, your work has rebuked me." With that, he thanked the group, and sat down, his "speech" having exceeded the length of its title, but not by much.

J. Waldo Smith, who was supervising the building of New York City's 163-mile-long Catskill Aqueduct, came out for a look at what Mulholland was doing and confided to reporter Kelly that he had expected little. Though he had heard something of the project, Smith attributed most of it to California "brag and boom talk." After his inspection, however, Smith admitted to being he saw, especially the courage of the men undertaking such work through such a forbidding region.

In the final analysis, and as an engineer well accustomed to the Tammany Hall quid pro quo system of inducements, Smith was most profoundly impressed by the fact that a municipality was handling the work "with its own force instead of by contract and doing it economically, efficiently and without graft or suspicion of graft." In Mulholland, Smith added, "they had the good fortune to find the right man and the good sense to trust him."

FAIR MONETARY RECOGNITION

ON OCTOBER 18, 1909, WITH WORK APPARENTLY going swimmingly on the aqueduct, the Board of Water Commissioners met and voted to raise Mulholland's salary, which had climbed to $10,000 over the years, to $15,000, bringing it equal with that of J. Waldo Smith, chief engineer of the Catskills water project. As one editorial noted, it was recognition of the services Mulholland had provided both as superintendent of the water system and as chief engineer of the aqueduct. His counterparts involved in similar projects elsewhere "usually command much greater salaries," the writer contended, and the raise was "only a fair monetary recognition of the services he is performing for the city."

In early November 1909, Mulholland spoke at the University of Southern California to announce that work on the aqueduct was virtually "half finished." In terms of mileage, the Chief clarified, only a quarter of the 233 miles were completed, but given that the most difficult work, that of tunneling, was almost finished, half of the work was behind him.

Mulholland did not tell the story on that day, but one notable incident had taken place during the summer's tunnel work on the Jawbone Division. Mulholland had come up the line for a consultation with division supervisor A. C. Hansen, an able engineer and manager, but well known as a man lacking in humor. Mulholland listened patiently as the dour Swede ran down the list of operations, drilling rate, and progress, then finally interrupted.

"How about the Pinto Tunnel?" Mulholland asked.

Hansen glanced up from his figures, clearly discomfited. "The Pinto's fine," he said. Then after some paper shuffling and throat clearing, he added: "Though we do have a man caught in there."

Mulholland stared back. "Is he dead?" the Chief asked.

"No," Hansen said, explaining that the man had been cut off from the surface by a cave-in. They were hard at work at the very moment boring a tunnel that would reach him. Meantime, crews had managed to drive a two-inch pipe through the muck that carried fresh air down to the trapped miner while they worked. "We've been talking to him through it," Hansen said.

"How long has this been?" Mulholland asked.

"Three days," Hansen said.

Mulholland, who'd heard nothing of all this, was astonished. "He must be nearly starved to death," the Chief said.

Actually he wasn't, Hansen said. "We've been rolling hard boiled eggs to him down the pipe."

Mulholland considered this with the same apparent gravity Hansen had used to deliver the news. "Well then," Mulholland asked finally, "have you been charging him for his board?"

"No," the concerned Hansen said. "Do you think I ought to?"

The unfortunate miner was eventually rescued, but thus was

born another story illustrating a legendary temperament. Mulholland was not a man who asked to be pleased by others; on the contrary, those who worked with him enjoyed his company and wanted to please *him*. A good part of it had to do with his pragmatic, get-the-job-done approach. As he once told John Gray, superintendent at the Elizabeth Tunnel, "I'd rather get a shovelful of muck out of the tunnel than all the cost reports on the job."

Whatever was to account for his ability to motivate men, the pace of the work continued to accelerate. On December 1, 1909, Mulholland appeared before a joint meeting of the City Council and the Board of Public Works to announce a formal revised completion date of May 1, 1912, a full year earlier than originally estimated. While he had surmised as much earlier, owing to the intricacies of the project's financing, Mulholland had waited to make a public announcement before the council.

In order to meet that revised deadline, more men would have to be put to work, and cement and steel and other supplies would have to arrive more quickly. Thus, the rate of bond sales would have to increase as well, with the entirety of what was left needing to be sold a year sooner than planned. This would mean paying out $250,000 more in interest in 1911, Mulholland pointed out, an amount that would have to be raised through an increase in taxes. In the end, the money would come back to the city in the form of revenues from water sales that would begin a year sooner than expected, but meantime, the city would have to authorize the added expenditure. And, more significantly, buyers for the bonds would need to be found quickly.

As difficult as it may be to believe from the present-day perspective, no one expected a great deal of resistance to this proposed tax hike. But the issue of bond sales was a more ticklish matter. By this time, William Mathews had resigned his post as city attorney to devote his full energies to the legal affairs of the aqueduct. His attempts to convince the Kountze-Leach syndicate in New York to accelerate any advances on the previously determined schedule of bond payouts fell on deaf ears. Bond sales in general had slowed, the investment bankers said; thus, they could not issue advances for sales they saw little likelihood of making any time soon.

As a result, Mulholland was not only unable to add to the work force—but he also faced the agonizing prospect of laying off men who were making such remarkable progress. There were 3,600 men at work through the first quarter of 1910, but by May, the money to pay them would be gone. According to his reckoning, 2,700 men would have to be let go. In desperation, Mulholland traveled to New York with Mathews in April, but even the Chief couldn't change the minds of the bankers, and with the work stymied, the dreaded layoffs had to be made.

Andrew Carnegie had recently appeared at a Los Angeles Chamber of Commerce banquet where Adna Chaffee announced his intention to see a bronze tablet erected in recognition of Mulholland's efforts, and when Carnegie, then seventy-four, rose to speak, he echoed Mulholland's frustrations with a financial system that seemed at times to be opposed to the very notion of progress. Carnegie, the steel baron who had always railed against companies whose chief product seemed to be the issue of stock certificates, told the Chamber of Commerce, "The men at the head of things, such as railroads and other great corporations—the real 'doers'—are poor men." Carnegie proclaimed, "In time the stock gambler and the parasites of Wall Street and their ilk will not be recognized as men by men."

It was a sentiment that Mulholland shared. Though he could not know the true reasons for the failure of those in what he had come to call "the halls of Mammon" to advance funds to the city, it was his certain feeling that the inaction was "calculated to seriously embarrass the city in its efforts to keep construction work going on the aqueduct."

Upon his return from New York, a frustrated Mulholland laid out a dire scenario: "We will have to do something immediately. . . . We must shut down all work and discharge all of the men employed on the aqueduct except for a few . . . or the banks of Los Angeles will have to invest in our bonds. We have a vast amount of expensive machinery which would stand idle; we have about 600 head of horses which must be fed, and we have a large amount of equipment representing considerable daily expense, whether idle or in use."

Los Angeles Mayor George Alexander opined that local power companies, backed by Wall Street interests, had worked behind the scenes to quash the advances on bond sales, but in the end it seemed that Mulholland and Mathews had been more convincing with the Eastern interests than they first believed. On July 19, papers, including the *Los Angeles Herald,* carried the news that the Kountze-Leach syndicate had found buyers after all: the New York Life Insurance Company was taking $500,000 of the bonds, and the Metropolitan Life Insurance Company was buying another $500,000. In addition, the syndicate itself was purchasing $530,000. Finally, with the coffers refilled, Mulholland was able to put 2,000 laid-off men back to work.

A reporter once declared that "if Mulholland told people he was building the aqueduct out of green cheese, they'd not only believe him but take an oath it was so." If it was overstatement, it was only moderately so, for ultimately it was the demonstrated progress on the route as well as the sound financial track record of the water department that convinced the Eastern interests that the aqueduct bonds constituted one "California boom and brag" investment that was sound.

Shortly after Mathews and Mulholland returned from their unsuccessful trip to Wall Street, the *National Geographic* published a long article in its July 1910 issue, accompanied by a number of dramatic photographs, highlighting the progress on the project—one hundred miles of aqueduct completed by June 1, with thirty-six of them tunneled, it was claimed—emphasizing the fact that the city, with Mulholland at the helm, was in control of its own destiny. Given the efficiency with which the work had advanced to date, there was no reason to doubt that it would be completed in accordance with the revised schedule and would begin paying for itself within two years.

"This is a public work without any politics," the piece pointed out. "There are no men on the payrolls who have outlived their usefulness, or have been failures in life and have found a berth because of friendship at the city hall. . . . Every man in a position above that of day laborer received his certification from the City Civil Service Commission."

The city's average daily consumption presently stood at about 35 million gallons, the piece went on to say, and it was projected that the figure would rise to about 110 million by 1925. The new aqueduct would deliver 260 million gallons per day all by itself. In addition, the story pointed out that voters had earlier in the year approved an additional bond issue of $3.5 million to be used for development of powerhouses along the route of the aqueduct in the San Gabriel Mountains. Mulholland estimated that power sales alone could net the city about $1.4 million per year. The added cost was well worth it, he said, an investment that would within twenty years "turn back into the city treasury the entire $24.5 million provided for the construction of the aqueduct, with interest."

Over the previous decade, the total manufacturing output of Los Angeles had grown from $5 million in 1900, to $30 million in 1905, and was predicted to top $75 million by the end of 1910. Given a burgeoning cotton-growing industry in the Imperial Valley, the outlook for milling operations in Los Angeles seemed bright, the piece asserted. Though the story was written by Burt Heinly, Mulholland's secretary, the statistics seemed to speak for themselves. What was costing Los Angeles $23.5 million, Heinly insisted, would have required $40 million if contractors were doing the work, and that presumed the unlikely prospect of no contractor cost overruns—none of which had been encountered thus far on the project.

Whether it was the persuasiveness of Mulholland, Mathews, or such encomiums, including those of Catskills project chief J. Waldo Smith, or Panama Canal engineer John Freeman, another $1.5 million had flowed in, and work was back on track. Though some of the 2,000 who were laid off had drifted to other work, a raise to $3 per day in the miners' pay had many skilled workers hurrying back eagerly.

On the night of October 1, 1910, however, the attention of the city was diverted from aqueduct issues, when the notoriously antilabor *Los Angeles Times* was bombed during a campaign to organize local workers by the International Ironworkers Union. Since 1906, the union had carried out a series of more than a hundred such bombings at steel works around the nation, intending to force management to the bargaining table. Though, as was usual, the saboteurs' intent was to be more provocative than deadly, they inadvertently placed their dynamite charge in a suitcase in an alley outside the *Times* building just above a spot where the gas mains ran. In addition, and unknown to the bombers, the building was full of employees hard at work on an Extra edition.

The explosion took out one wall of the building, and the resulting fire destroyed it and the neighboring printing plant, killing twenty-one inside the building and injuring more than a hundred. Though the Ironworkers Union and Samuel Gompers, head of the American Federation of Labor, immediately denounced the action and disavowed any connection with it, most assumed that labor organizers were responsible.

The furor delayed an excursion of Chamber of Commerce and City of Los Angeles officials intended to mark the virtual completion of the Southern Pacific's line through the Owens Valley, but, finally, on a Saturday morning, October 29, Mulholland and Mayor Alexander led a procession of twenty-five automobiles northward out of Los Angeles. The group stopped for a tour of the works at the South Portal of the Elizabeth Tunnel and pressed on through the San Gabriel Mountains, with Mulholland at times guiding the motorcade in surreal detours through some of the tunnels that would one day be full of aqueduct water. By nightfall, the group reached Mojave, where, given that there were nearly a hundred people in the party, they were put up in a series of Pullman cars on a rail siding instead of at the Harvey House.

The following day, the party was off from Mojave, bound for the foot of the Owens Valley, where they would spend the second night. Along the way, Mulholland diverted the procession to view a section of covered aqueduct crossing the upper stretches of the Antelope Valley. From that point all the way to Haiwee Pass, a distance of nearly a hundred forlorn miles, there remained only twenty-three miles of aqueduct to be completed, Mulholland told them.

Near the Red Rock Summit, Mulholland led the group on a hike up to a spot that he told them he had named "Point Despair." During the original surveying, he said, his confidence was shaken when he got to this rugged division between the Salt Wells Valley and the Antelope Valley below. The drop in elevation there was so steep that he could never get the water up and over the ensuing side, he said, with a wave over the vast declivity. The only way to maintain the necessary elevation was to go through *that,* Mulholland said, pointing at a sheer cliff to their west. It was the place at which the series of unprecedented tunnels and siphons of the Jawbone Division had begun. Moments later, Mulholland had the party clambering down a steep trail and into a cool bore that now pierced the way completely through the rock.

That night, at Haiwee, Fred Eaton joined the party, having journeyed down the Owens Valley on the new rail line, and he and Mulholland regaled the dinner party with tales of their original trip over the route and their eventual agreement that the impossible thing could in fact be done. Finally, Eaton slipped away, bound on the night train back to Independence where he would ready a party for the group on the third night of their journey. The others sat on boxes and barrels in the aqueduct's offices, listening to yet more tales, including one by former Los Angeles sheriff Billy Rowland, the man who captured the legendary Tiburcio Vasquez, a bandit who preyed on miners plying the lonely route down from the Nevada gold mines.

As Rowland explained to the group, they "might none of them be there" that evening if it were not for the legendary luck of the Irish. Back in his days as a ditch tender for the water company, the sheriff explained, Mulholland was returning to his shack beside the Los Angeles River below the Glendale Narrows one night when he came across a group of rustlers who were slaughtering a stolen cow. Mulholland had the good sense to simply wave and say "Hello, boys," as if nothing were wrong.

He rode on to his cabin, went inside, and lit a lantern to make it appear to anyone who could have followed that he was getting ready for bed. After a few moments, he slipped out the back door and walked all the way to downtown Los Angeles to tell Rowland he'd spotted the group suspected of a series of cattle thefts in the area. By the time Rowland and his men got back to the spot, though, there was little left but skin and hooves.

The following morning, the sheriff was back, passing by Mulholland's cabin with his posse. Mulholland saddled up and rode along to watch. Not far from Elysian Park, the posse spotted the rustlers on their way toward the wastes of San Fernando Valley, and suddenly the chase was on.

Mulholland's horse was a good one, Rowland said, maybe too good. Soon the mare had outpaced the sheriff and his posse, and Mulholland was helpless to rein her in. Some of the thieves turned to fire at their pursuers, and members of the posse began firing back. An unarmed Mulholland, caught in the middle, flung his arms around his mare's neck and held on for dear life while the shots whizzed back and forth. Finally, one of the sheriff's men shot a rustler dead off his horse and the others gave up the game. The long and short of it was, as Rowland assured his listeners, Bill Mulholland was no cattle thief but he damn near got shot for one. On that note, the party retired to bed.

The tour was to conclude with a visit to the Cottonwood Power Plant the next day, followed by ensuing nights in Independence and Bishop where citizens greeted the party with enthusiasm, according to reporters, going so far as to serve up a dinner of wild game at a local club on the last evening. By the end, the trip had covered 650 miles and made a firm supporter for the project out of Mayor Alexander, who assured reporters that he had found "the work of building the aqueduct all it has been reported to be." One event brought a rebuke from Mulholland upon his return. On November 1, it had been reported that a general strike was underway by workers at both the North and South Portals of the Elizabeth Tunnels as a protest against the hike of five cents in meal prices—from twenty-five cents to thirty—at the Desmond commissaries. The food was bad enough at the former price, many said. Until their pay was raised accordingly, sources said, the men would stay out.

Mulholland dismissed the allegations of any organized unrest, telling reporters that he had spoken to the superintendents at both stations "not two hours ago." He said that while some 200 men had left the 3,000-plus workforce over the last few weeks, it was not an unusual number. "When cold weather begins to approach, a number of them prepare to go elsewhere." Mayor Alexander joined in the rebuttal, assuring reporters that he had seen no evidence of any strike or protest during his tour.

Given the restiveness created by the still-unsolved bombing of the *Times*, and the efforts by unions to make headway in what had always been an open-shop city, speculation as to possible labor troubles in any arena was at a peak. But Mulholland was confident that the aqueduct pay scale was more than adequate to maintain an eager workforce despite any union efforts. Laborers were being paid $2.00 to $2.50 per day and an additional $1.00 for their meals; miners were making $3.50 per day plus board and could add as much as 40 percent to their base salaries with bonuses. In the current economy, those were generous figures.

The pay scale published by the Board of Public Works in 1911 listed the pay of blacksmiths and carpenters at $3.00 to $4.00 per day, shift bosses at $3.50 to $4.00, concrete workers at $3.50 per day or $160 per month, electricians at up to $175 per month, clerks starting at $70, shovel operators at $130, and engineers ranging from $100 all the way to $833 per month.

Of more importance to Mulholland was the need to raise the $1.25 million necessary to pay for the steel pipe to complete the siphons. To do that, the rate of bond sales would have to pick up, but, as news reports laid out, in order to induce the New York syndicate to revise its contract and exercise its option on the nearly $3 million in bonds remaining during the calendar year of 1911, the city would have to relinquish about $62,000 in premiums it was due under the current agreement.

Not surprisingly, the City Council voted in accordance with Mulholland's recommendation, but the following day Mayor Alexander announced that he had personal reservations about the deal. Not only would the city lose out on the premiums previously negotiated, he said, it would also have to agree to various restrictions on the use and accounting of funds, matters that the mayor considered of significant detriment.

Meantime, confirmation of organized stoppages along the line arrived, with 350 men reported to be on strike at various locations, though work was not significantly impeded at any division except Haiwee, where the miners involved in the extraction of tufa material used for cement making had shut that operation down. According to the division chief there, the stoppage was in fact working to the project's advantage, as they had far more of the volcanic rock piled up than could presently be used. In addition, miners who had been ordered off by their unions were reported to have returned to the job as laborers, and Mulholland seemed unconcerned for the moment. He told reporters that he would simply use any savings from payroll to put some of the idled steam shovels back to work.

FITS AND STARTS

IN THE END, THE FUNDING SLOWDOWN ACTUALLY WORKED in Mulholland's favor. With much of the hardest tasks behind him, any walkouts simply meant fewer men to pay, and he was confident that once funding was fully restored to the project, he could put enough men to work to bring the aqueduct in on time and under budget.

Joe Desmond's operations had long been the subject of workers' complaints, though even if he had been a more experienced commissary man, supplying fresh food over such distances in those days presented great challenges. As young laborer Erwin Widney put it, "When you are in the desert, 150 miles from the producer, with no ice, no fresh vegetables or fruit, no butter, eggs or fresh milk, you are somewhat limited in your bill of fare."

Still, the cooks, at least in the independent camps, did their best. Principal items on the menu at Camp 30-A included "fresh meat which had not lost its body heat, beans, much gravy, potatoes, onions, bread, stewed raisins or prunes, pie, and coffee and tea." The latter two, Widney claimed, were indistinguishable, "except for the red string tied to the coffee pot." Likewise, Widney found little to appreciate in the ambiance surrounding mealtime, for as he put it, "Eating was a duty, not a social event," with little of substance passing between himself and his fellow diners beyond the occasional "Let's have the potatoes."

Surveyor's helper Frederick Cross also remembered the food in the smaller crew tents as being good, but at Desmond's mess in Cinco, the food was "execrable." He recalls that two enterprising Frenchmen at one point brought a tent onto unspoken-for lands in the desert near the camp and set up a veritable dining emporium, giving workmen "an opportunity to rejuvenate their stomachs." However, when Desmond learned of the competition, he had the workforce reminded in no uncertain terms that subsistence would be deducted from everyone's pay whether they took their meals in his tents or not.

Cross also remembers that one day Desmond had made the "cardinal error" of having himself chauffeured into the camp at Cinco in an elegant new Mitchell touring automobile just as the men were changing shifts. The result, Cross says, "was a volley of abuse as only Irishmen know how to direct."

Not the least of Desmond's difficulties involved recruiting able cooks and keeping them on the job under trying conditions. One of the standing jokes was that it took three crews of cooks and helpers to man every mess house: one working, one coming up the line to go to work, and one on the way home.

As he traveled to his first assignment on the aqueduct, Widney was in fact accompanied by one of Desmond's crew chiefs who was escorting four new cooks bound for work in the commissaries. These were the toughest cooks in the cooking business, the crew chief announced, but he was going to see that they made it to their assigned stations if he had to kill them all with his bare hands.

"Each had a quart bottle of whiskey," Widney said of the cooks, describing them as "the hardest boiled characters unhung." But he admitted that they could not be said to have been drinking. More precisely, he said, "They merely put the neck of the bottle down their throats and the whiskey flowed down without a swallow."

Medical Services director Raymond Taylor had been doing what he could to encourage Desmond to upgrade the food, but even he admitted that his fellow contractor was facing a difficult task. For one thing, "There was no ice to be had," Taylor said, "and shipping ice from Mojave by train or trying to get it out by automobile was a pretty poor proposition . . . if he sent it on the train . . . two-thirds would melt or be stolen before it got to its destination."

Of equal concern to Taylor was the question of sanitation, including the near-futile attempt to keep flies out of the latrines, cookhouses, and mess halls. Screens were employed, and helped, but doors were always being propped open and panels kicked out, and the ubiquitous presence of the mules ensured a chronic problem. After some effort, Taylor managed to see that the corrals were situated downwind from the cookhouses, a practice that he said seemed groundbreaking to many of the camp chiefs.

As for the problem in the latrines, Taylor came up with another inventive solution. He had observed that flies do not generally congregate in dark places and so convinced the camp engineers to erect privies with doors facing south, and baffles creating a kind of two-turn maze at the entrances. It left men enough light to get inside and find the toilets, but the problem of flies was greatly decreased. In the end, Taylor said, only one outbreak of typhoid struck the workforce in the years during construction. Though fourteen men were diagnosed, all of them were cared for at the hospital in Cinco, and no case proved fatal.

The year 1911 began auspiciously with a *Los Angeles Times* story on January 12 declaring, "The outlook on the Los Angeles aqueduct was never brighter." The City Council had appropriated $500,000 of its own monies to circumvent the flap over the bond renegotiations, and the strike seemed a thing of the past. There remained only 1,295 feet of rock to be dug to complete the tunnel beneath Lake Elizabeth, and Mulholland's continual innovations suggested that he should consider turning author as soon as he finished with the job that had caught widespread attention.

Among other things, Mulholland had been the one to suggest that the volcanic tufa deposits on lands adjacent to the rail line at Haiwee could be ground up and mixed half-and-half with more expensive Portland cement being manufactured at the city's plant near Mojave. While it took a bit longer for the mixture to set up, the end result—a form of which had been used by the Romans in their road and aqueduct building—was even stronger than cement alone, Mulholland insisted. Furthermore, since it cost a dollar a barrel to haul cement from the plant to work sites up the valley, using the tufa already in place at Haiwee to mix with the cement cut the freight cost in half. All in all, Mulholland said, tufa mining would save $200,000 before the job was done.

More notably, the *Times* reported, the big steam shovel that had been operating in the Alabama Hills above Olancha was no longer a "steam" shovel at all. The pistons of the machine were now being driven by compressed air, generated by free power from the project's station at Cottonwood Creek. In addition, the dam at Haiwee was now excavated down to bedrock, and Mulholland announced that much of the dam's core would be formed by a novel process whereby clay and dirt would be forced into place through the use of hydraulic jets instead of laborious hauling and filling.

Finding ways to cut waste and increase the efficiency of his crews had become a fixation with Mulholland. Once he was called away briefly to testify before a hearing of the State Railroad Commission regarding the application of a Marin County water company to increase its rates. The hike was necessary, the company claimed, owing to the high cost projected in building a new earthen dam.

During cross-examination, the company's attorney approached Mulholland in the witness box and asked, "Now, Mr. Mulholland, considering the location of this dam and all the surrounding difficulties of construction, would you not think it might reasonably have cost 80 cents a yard?"

Mulholland paused, apparently giving the question careful consideration. Finally, he leaned forward. "Well," he said, "if you had had a parcel of old women, carrying that dirt up on the dam in their aprons and stomping it down with their feet, it might have cost 80 cents per cubic yard."

As for the present aqueduct project, the *Times* assured readers, the situation was sanguine. "Money and men, with plenty of both in sight, are all that is required to complete one of the most gigantic undertakings of a modern city."

At the beginning of 1911, only 1,300 feet separated the two crews working beneath Lake Elizabeth, and the principal part of the tunneling through the Jawbone and Grapevine Divisions was finished. Much of the steel siphon work in both rugged divisions remained to be done, owing to the expense of the pipe and the necessity for it to be ordered about a year ahead of its actual placement in the field. Nonetheless, Mulholland had nearly $1 million in the coffers, almost enough to get through the coming year, with close to $3 million of bonds left to sell.

The cutbacks of the previous summer had made it necessary to shut down some of the big steam shovels along the route, but a number of these machines were still at work on the Olancha Division near Haiwee and at other points along the Mojave portion of the aqueduct. For many of those who had gone on the trip, one of the highlights of the previous fall's tour—in addition to the thrill of driving an automobile through part of the pipeline—was the opportunity to watch the relatively unfamiliar machines at work. There had been steam shovels before, but Mulholland was responsible for the size of these beasts, as well as a number of advancements in their technology.

When the board first advertised for bids for machines that could operate according to Mulholland's specifications, most manufacturers regarded the standards as impractical and refused to submit proposals. Mulholland finally met with representatives of the Marion Steam Shovel Company, an Ohio company that had been in the business since the mid-1880s. Engineers for the company told Mulholland that they would build machines of the size and power he stipulated, but they couldn't guarantee that they could actually do the amount of work he expected.

Mulholland's response was what one could expect. "You build the shovels," he said. "I'll take care of the guarantee." In the end, the company went along, the shovels worked as efficiently as Mulholland intended them to, and soon the company had placed several similar machines at work in the digging of the Panama Canal, as it would in many vast earthmoving projects to come. It was just one more instance of Mulholland's work making its permanent mark far beyond the bounds of his adopted home.

While papers carried stories that steam shovel operators affiliated with the Western Federation of Mines walked off the job early in February, they were quickly replaced by non-union men. Given the far-flung, decentralized nature of the work along the aqueduct, it was nearly impossible to organize a picket line, or for that matter, to get union organizers into the camps at all.

Meanwhile, the shovel operators' strike was overshadowed by the news that on February 27, 1911, tunnel crews finally broke through the rock to meet at a point about 250 feet beneath Lake Elizabeth after forty months of work, two-thirds of the time initially projected. Crews had driven 13,500 feet from the South Portal and 13,370 feet from the North, and might have met considerably earlier except for the fact that at 1,117 feet inside the mountain, miners blasted into an ancient fissure filled with sand and water.

The mishap sent men scrambling back up the tunnel; fortunately, no lives were lost. The hole had to be plugged, however, the water pumped out, and a method determined for getting around the roadblock. Finally, Mulholland ordered a shaft driven down some 3,000 feet from above, culminating in a spot beyond where the fissure ran. From there, miners inched their way back northward toward the fissure, driving overlapping steel plates to support the tunnel ceiling as they went. Finally, they were able to run their sheeting through the fissure to the place they had reached before the blowout, and progress could begin again.

When the crews finally met near the midpoint of the five-mile project, Mulholland said, "The center line of the tunnel met within 1⅛ inches and the grade check within ⅝ inch," quite the accomplishment for such a project in that day. Furthermore, while the average rate of progress was projected to be eight feet per day at the outset, his crews, driven by the bonus schedule, had averaged more than eleven feet per day at a savings to the project of half a million dollars.

With what Mulholland called "the controlling event" of the work behind him, it seemed that there was little of substance left standing in the way. Of labor matters, he had little to say in his "Sixth Annual Report," issued in July, beyond an observation that labor conditions had not been "entirely favorable" during the previous year. Mulholland offered only a sketch of what he had encountered from the outset: "In the summer, when the weather is hot on the desert, the laborers make a general migration to the northern portion of the United States and British Columbia, where work is abundant at that season, under conditions that are more agreeable. As the winter months approach, this labor largely returns and there is a plentiful supply until about the first of May, when the northern exodus again starts. Consequently, the winter months are the best months for vigorously prosecuting the work."

He lamented the fact of the enforced slowdown in 1910, when the work force had to be cut in half. Overhead on the project remained relatively constant, he pointed out, whether there were 1,500 in the field or 3,000. Still, he said, he hoped to complete the aqueduct by the spring of 1913. Technically, Mulholland was now reporting to new superiors at least half of the time, given that an amendment to the city charter in March abolished the former water board and replaced it with a new Public Service Commission.

The new body took charge of a water works system that took in more than $1 million, with per capita water consumption reduced from 306 gallons to 140 gallons per day, and rates to consumers less than half of what they had been at the outset of the city's takeover in 1902. The group was also charged by Mayor George Alexander with developing a new bureau of power and light with Ezra Scattergood as chief. Mulholland would remain in charge of all things water related, while Scattergood would see to the distribution of power generated by the new stations to be constructed along the line of the aqueduct in San Francisquito Canyon. So far as aqueduct construction was concerned, however, Mulholland remained the Chief. Mulholland advised in his July report that work in the San Francisquito Canyon was delayed owing to various legal actions undertaken by the private power companies opposed to the city's entry into the business, and he also noted that the last remaining work of significance lay in the Jawbone and Grapevine Divisions where the siphons remained to be finished.

If the physics of the project seemed to have bent to Mulholland's will, however, the ensuing mayoral campaign would soon prove that metaphysics was far less compliant. The machinations of Harrison Gray Otis and Harry Chandler, Otis's son-in-law and heir apparent at the *Times*, in opposition to various labor movements of the era are labyrinthine and have formed the stuff of numerous publications. By 1911, however, the influence of the American Federation of Labor and the Socialist Party had created a significant change in the political landscape of Los Angeles, where business interests and Republicans had held sway.

While Mulholland had always enjoyed an unqualified reputation as an apolitical "man of the people," that reputation was about to be called into question. Job Harriman, a vice-presidential candidate on the Socialist ticket in 1900, and Socialist nominee for mayor of Los Angeles for the December 5, 1911, election, had been an opponent of the aqueduct from the beginning, arguing that it was ill designed, incapable of delivering what was promised, and unnecessary in light of the copious flow of the Los Angeles River.

With the aqueduct only about fifty miles shy of completion and the population of the city verging on 350,000, most of Harriman's criticisms had been disposed of. However, there was one damning issue that remained, and he seized upon it vigorously in his campaign against the incumbent, Mayor Alexander. The aqueduct had been promoted for only one reason in the first place, Harriman claimed: it was never intended to benefit the citizens of Los Angeles, only a much smaller group, the members of the San Fernando Land Syndicate.

There was little he could do to put a stop to the aqueduct at this point, Harriman declared, but given that the city was prohibited by state law from selling Owens Valley water to any individuals outside its limits, he would dedicate himself as mayor to ensuring that the manipulators behind the project would never receive a drop. For his part, General Harrison Gray Otis began a publicity campaign not so much in Alexander's favor but as a merciless smear upon the "anarchic scum" represented in Harriman's candidacy.

In the meantime, in April 1911, two brothers and union operatives, James and John McNamara, had been arrested in Detroit, charged with the October 1910 bombing of the *Times*. The pair were extradited to stand trial in Los Angeles, though a number of labor supporters contended that the brothers had been grilled and tortured by a private detective for days inside the home of a Chicago police chief before being virtually kidnapped to Los Angeles. The McNamaras had been set up by Otis and his cohorts, many claimed; Eugene Debs, founder of the Industrial Workers of the World and long-time Socialist leader, went so far as to speculate that Otis had arranged for the bombing of his own building so that he could discredit the union movement in Los Angeles.

Union leaders of every stripe banded together to raise funds for the McNamaras' defense and in the end, Clarence Darrow was retained as defense attorney, with none other than mayoral candidate Job Harriman appointed to his team. The lead-up to the trial, with labor supporters decrying Otis and idolizing Harriman, seemed to promise an end to the days of Los Angeles as an open-shop city. On the other hand, Darrow had become increasingly concerned that the McNamaras could in fact be proved guilty. Popular muckraking journalist Lincoln Steffens, in town to cover the trial, had spoken with the McNamaras and had gone to Darrow to share his concerns. The pair might indeed have been treated shabbily by authorities, but Steffens thought that they had probably done what they were accused of.

Steffens had an idea, though. Perhaps a plea-bargain deal could be brokered. The McNamaras would be spared hanging in return for light sentences, and Otis would agree to reopen negotiations with typographers and other groups wishing to unionize within the *Times*. Otis scoffed at the idea of negotiating with the unions, and wanted nothing more than to see the McNamaras hanged, but his son-in-law Harry Chandler argued that a plea-bargained admission of guilt would achieve the greater aim of discrediting the union movement among Angelenos. Job Harriman, meantime, knew nothing of the negotiations.

Finally, on December 1, just four days short of the mayoral election, the bottom fell out for Harriman. Clarence Darrow stood up in court to announce that the brothers McNamara had changed their plea to guilty. John McNamara admitted that he had placed the suitcase full of dynamite outside the *Times* building on the fateful night. Harriman, who read of the admission in the newspapers, was stunned. Hours before, he had seemed a shoo-in for mayor. Now his candidacy was doomed. On December 5, he was defeated by a margin of some 34,000 votes out of the 137,000 cast.

Though it would have been precious little consolation at the time, Darrow later explained why he had blindsided Harriman. The lives of his clients were at stake, he wrote in his autobiography. "I had no right or inclination to consider anything but them." If Darrow had advised Harriman of what was being considered, it would have placed him in the position of either quitting the race or sending at least one of their clients to the gallows, a choice that Darrow said he wished to spare him.

FALLOUT

THE MCNAMARAS' CONFESSION TO WHAT HAD BEEN termed "the crime of the Century" constituted a watershed moment in American labor history, as devastating as the quashing of the Homestead Steel Strike had been in Pennsylvania in 1892. In the same way that it would take labor organization within the steel industry four decades to rebound from the Homestead calamity, it would likewise take forty years and the unprecedented manufacturing boom of the post–World War II era before the open shop in Los Angeles was challenged again.

Mulholland was no outspoken opponent of labor, save for his concern that he be able to complete the aqueduct for the sum that he said it would require. In fact, he lent his stated support, along with that of a number of other prominent members of the Los Angeles business community, to the compromise that Lincoln Steffens had tried to broker between Harrison Gray Otis and Clarence Darrow. Steffens later wrote that he was happy to have the support of reasonable men such as Mulholland, but that his ideal had been to bring polarizing figures such as Otis and Eugene Debs into conversation together. As for the likes of William Mulholland, "he was too public-spirited" to be useful for Steffens's quixotic purposes.

Nonetheless, Mulholland was stung by Job Harriman's renewed criticism of the aqueduct project, and on November 27 he delivered a spirited defense before a City Women's Club luncheon of 600, saying that he had had his fill of misrepresentation and distortion prompted by nothing but political gain. Eighty percent of the project was complete, he said, and it had taken exactly 80 percent of the funds to do it. "The other 20 percent will be built with the other 20 percent of the money," he told his audience.

As to the claim that the 20,000 inches of water brought from the Owens River were far in excess of the city's present needs, and the suggestion that the excess should thus be left in the Owens Valley, Mulholland scoffed. He admitted that he too had once been doubtful of the need for more water—he had even told Fred Eaton as much: "Not long ago, I was a pessimist," he said. "I believed that growth would stop and we would have a chance to catch up in the way of water."

But time had proved him to be shortsighted, Mulholland said. Each year the city was demanding an additional 250 inches of water and there was no end in sight. "Are we to be fools and only take that which we need from year to year and leave the rest up in the Owens Valley? Are we to lose the great electrical energy the full quantity of water will produce?" The full flow of the aqueduct was required to generate power, and while that flow would indeed be in excess of the city's present needs, that water was the property of the City of Los Angeles and it would not be wasted but sold in order to help recoup the project's cost.

As to where the excess should be sold, that was a question for voters to decide, Mulholland said, though he did give his view of the controversy surrounding the San Fernando syndicate. "Some say it must not be sold to the San Fernando Valley because a syndicate owns a lot of the land. Well, if you sell it to Cahuenga or the Redondo region, you will find that the land there is owned by somebody. In fact, anywhere you put it someone owns the land."

Mulholland also responded to those who criticized his proposal for building impoundment reservoirs for the Owens River water in the San Fernando Valley. "There is a loud wail because the aqueduct stops at the head of the San Fernando Valley or that it comes by way of the valley." He paused with characteristic Mulholland timing to glance around the room. "Well, we couldn't bring it by way of Catalina and San Pedro, nor by way of San Bernardino." The route through the valley was the most direct, he reminded them, and it crossed the narrowest part of the intervening Coastal Range to get there.

Nor did the water stop there in the valley reservoirs, he said, pointing to a slide projected onto the wall behind him. "That black line you see coming down (from the Valley Reservoirs) is the route of the conduit the Public Service Commission is ready to build to bring the water to the city." That conduit would use public roadways for its route, he said, and its size would be determined by the amount of water that voters wished to sell and by the location of the lands consuming it. In any case, that conduit would bring water "ample for all city purposes for decades to come." Furthermore, he said, the conduit was being brought down the west side of the city's present limits for a simple reason: "Because the growth of the city is to the west."

Mulholland's contention on the latter score was anathema to Pasadena and other eastern communities that hoped to avail themselves of the excess water, but part of his reasoning was grounded in the fact that supplying water to unincorporated potential locations at higher elevations in the eastern part of the county (Azusa, Glendora, and Covina, for example) would require costly pumping and piping arrangements. Even allocations to Pasadena would be "water down the drain," so to speak, for engineers calculated that fully one-quarter of any irrigation waters supplied to San Fernando Valley lands would eventually percolate down into the Los Angeles River aquifer and return to the city's supply. However, waters supplied east of the Glendale Narrows would not reenter that same aquifer.

That debate was a fine point, and a matter for another day. He closed his talk with a swipe at those who invented conspiracies for their own ends, lambasting one person who had written to the bond-buying syndicate of Kountze-Leach complaining of egregious cost overruns on the project. A recent break in the line near the intake gate above Independence had cost $360,000 to repair, the complainant charged. In fact, Mulholland told his audience, the repairs had cost $360. Perhaps the letter writer had just made an honest mistake?

A week later, just prior to the election, Mulholland issued a similar rejoinder to Harriman before the City Men's Club, with particular emphasis upon charges that he had concocted a "water famine" in order to stir up support for an unnecessary project. He recited a battery of US government figures on the flow of the Los Angeles River for the period and also addressed another old charge: "They say we emptied the reservoirs into the outfall sewer," Mulholland said, referring to the 1904 incident that would later be recast as part of the plot of *Chinatown*, "but I need only remind you that during this water famine, when we had so much to waste, the superintendent of streets tried to have me arrested for turning the water off of the city sewer flush tanks."

On December 9, the week after Harriman went down to defeat by Mayor Alexander, Mulholland, still rankled by the suggestion that he had any illicit motivations in undertaking the project, sent a letter to his supervisors at the Board of Public Works, requesting that the board press the City Council to appoint "a committee, composed of its own members or private citizens, as may be deemed best, to make, with independent engineering assistance, a thorough investigation, not alone of the physical features of this work, but of the field administration and all other departments thereof as well." It was his wish, Mulholland said, that the citizens thus be assured of "the true conditions of this work . . . and judge for themselves whether it has been carried on in a proper manner." If Mulholland's intentions were in fact noble, it was an instance where a friend might have counseled him to leave well enough alone.

The City Commission, as could have been expected, responded as it usually did when prompted by Mulholland and quickly voted to appoint such a panel of experts. From that point forward, however, the matter deteriorated into the type of committee work sure to produce a camel. There was much discussion as to just how many members the investigating committee should have, and an equal amount of debate arose as to the ideal political constituency of the group.

By the end of January, a weary City Council finally ended debate with the Socialist opposition and named a three-person Aqueduct Investigation Board, to begin public hearings on February 7. Any citizen with relevant information or complaint was invited to come forward. Angry Socialists began a drive to force a voter initiative to reconstitute the committee to five members, with two Socialist Party members added, and to expand the scope of the investigation to include the original procurement of land and water rights and the nature of contracts made between the city and individuals or corporations. By the end of March, enough signatures had been collected to force a special election on the matter, scheduled for the end of May.

In the meantime, the chronic issue of funding had presented itself once again. On a crisp morning in early January, the city clerk opened a letter from the Kountze Brothers and Leach & Co. and stopped short at the terse message that ended, "owing to the position of the city of Los Angeles bonds, the syndicate decided today not to exercise its option under the terms of our contract." This was not a refusal to extend credit or accelerate the rate of purchase but a flat-out denial of interest in the remaining $4.2 million in bonds necessary to complete the project. Mayor Alexander, who had earlier in the week assured the City Council that exercise of the option was a foregone conclusion, was flabbergasted. Without the bond sales, there was only enough money left in the coffers to keep the work going for sixty days.

There was also great speculation as to the meaning of the phrase "owing to the position." Some interpreted that as a reflection of uncertainty regarding the growth of the city's indebtedness, as voters had approved both a $3.5 million bond issue to develop a municipal power system *and* another $3 million to finance improvements to the harbor at San Pedro. Others, however, attributed the turndown to the influence of the existing electrical power interests upon the financial houses, who were heavily invested in private power in the Los Angeles area.

Project attorney William Mathews was already on a train bound for New York, where he presumed he would be facilitating a simple process, and was himself dumbfounded when he found a telegram awaiting him with the news. Mathews was aware that Mulholland's patience with the bond-selling process was wearing thin, but there was little to do but persevere. After attempts to persuade Kountze-

Leach were fruitless, Mathews approached James Speyer, who had underwritten bonds for the original acquisition of the power company a decade before.

After some consideration, Speyer told Mathews that he would purchase all of the outstanding bonds—aqueduct, power, and harbor—under one condition: that the city would agree not to incur any additional indebtedness until January 1, 1913, a move that Speyer considered necessary to preserve the standing or "condition" on the bonds in Eastern financial circles. Furthermore, he was only buying about $2.9 million in aqueduct bonds—the $1.3 million that the city had placed in its sinking fund as a hedge would remain there.

The good news was that Speyer would make his payment on those bonds by March 11, 1912, thereby completing the entire $23 million issue at last. A relieved Mathews wired the news to Alexander, who called an emergency meeting of the City Council on the evening of February 10, where the measure was quickly adopted. The money was as good as in the bank. "LOS ANGELES' GREAT PROJECTS NOW FINANCED—TEN MILLIONS OF BONDS BRINGING QUICK CASH," headlines blared, and the completion of the aqueduct—assuming all went as planned—was assured.

Meanwhile, the three-member Aqueduct Investigation Board (AIB) appointed by the City Commission had begun its hearings, but the proceedings were almost immediately derailed when one member, Charles Warner, insisted on issuing a preliminary report to the City Council so that immediate steps could be taken to improve the food being served at Joe Desmond's cookhouses. Warner was particularly troubled by complaints that the condensed milk served to men was excessively diluted, that margarine was often substituted for butter, and that supervisors and special visitors were provided with better food than laborers.

Eventually, Mulholland and General Chaffee appeared before the Public Welfare Committee to defend the commissary system. Chaffee said that he was not quite clear on what the problem was with margarine: it was served on a regular basis to army and navy men, he said, and no one ever complained. Mulholland told the committee that he had often eaten at Desmond's camps alongside the men, eating the same fare that they had. It was the Chief's suspicion that most of the complaints about Desmond's mess came from people "who hadn't eaten there three times."

As the date for the special election neared, the AIB also heard complaints from representatives of the Portland Cement Manufacturers Association that the tufa cement manufactured by the city at its plant above Mojave was inferior, and another from W. T. Spilman, owner of a small San Fernando Valley water company that would likely be put out of business if aqueduct waters were made available in the region. Spilman, a critic of the project from the outset, had recently published a pamphlet, "The Conspiracy: An Exposure of the Owens River Water and San Fernando Land Frauds," whose subtitle conveyed the gist of the charges made. The aqueduct was nothing but a plot hatched to enrich Otis and others who had bought land for $50 an acre and were now selling it for twenty times as much. (Some lands had changed hands for as much as $350 an acre, though it would be quite a while before the figures claimed by Spilman were realized.)

When asked by the AIB to comment on the charges Spilman had made, Mulholland began by dismissing the tract as too hysterical to take seriously. In fact, he had been quoted by a reporter as excoriating "capitalists" who had possibly profited by speculating in agricultural land, but he was offended by suggestions that he had been in league with any such persons. Good sense, and nothing else, dictated that the water would have to come to Los Angeles through the San Fernando Valley. It was unfortunate that land that should have been used for agricultural purposes "has been subdivided into town lots and small rich man's country estates at prohibitive values," Mulholland said, but added, "I do not care whether the San Fernando Valley receives the water or not. I have merely done my duty as an engineer. . . . [The water] must be used somewhere."

As the three-member board appointed by the City Council struggled to come up with a consensus on these matters, the special election was held on May 28, and voters approved the reconstitution of the AIB to five members by a vote of 16,564 to 15,697, perhaps the first time that citizens of Los Angeles had gone against what Mulholland would have wanted (though it might be noted that he had not taken any active part in campaigning on the issue). On July 9, the new five-member board reconvened and began to hear the testimony of some sixty individuals who had been involved with the project, beginning with the question of how the city had initially identified the Owens River as the most likely source of future water.

With the two Socialists added to the panel, it would come as no surprise that charges leveled by Job Harriman would receive renewed attention. In "The Coming Victory," a pamphlet he issued during his mayoral campaign, Harriman laid out a "water plot" that played well to those segments of society perennially drawn to or eager to make use of what could be called the conspiracy theory of history.

> Big business, realizing the wonderful possibilities of profit to be made in exploiting land and water in the vicinity of Los Angeles, conceives a gigantic plan, and starts to carry it out with official aid. This plan involved the gobbling up of all available lands in and near San Fernando valley . . . [and] the

> securing of the Owens Valley water to irrigate these lands, by first creating a fake water famine and frightening the people into building an aqueduct . . . thereby putting about $50 million dollars [*sic*] profit into the corporation's pockets, while the city gets none of the aqueduct water.

Harriman was one of the few who went so far as to name the water superintendent as a coconspirator, claiming, "Fred Eaton goes to the Owens Valley and buys water rights; and Mulholland prepares the minds of the people with his reports of a 'water shortage' when there is an abundance."

Mulholland was, of course, among those called before the reconstituted board, and he began his testimony on Wednesday, July 10, the second day of hearings. After three days of listening to the superintendent answer virtually the same questions that had been put to him in the initial hearings, two of the original board members appointed by the City Council had heard enough. On July 15, Edward Cobb and Edward Johnson announced their resignation from the new board and followed that announcement by issuing their own report based on the hearings that the three-man panel had conducted prior to the May election.

Cobb and Johnson asserted that they had heard evidence on every salient issue, including the original purchase of Owens Valley lands, the activities of all land syndicates whose purchases had been called into question, the matter of distribution of excess aqueduct water, the quality of construction, and all other matters appertaining thereunto. As to their findings, they said, "There has not been brought to our attention one particle of evidence that would reflect in any way whatsoever upon the integrity of the management of the Aqueduct proposition from its inception to the present time."

While the resignations and report created something of a public stir, the reconstituted AIB, now a body of three, with two Socialists and one Progressive, Charles Werner, nonetheless soldiered on, calling Mulholland's chief assistant, J. B. Lippincott, to testify. Some opponents of the aqueduct considered Lippincott a far more despicable player than Fred Eaton in the so-called water plot. In their view, Lippincott had betrayed his public trust when, as a Reclamation Service official, he had tipped off Eaton as to the US government's intentions and aided Eaton in purchasing key land parcels that would forestall any reclamation project in the Owens Valley. Lippincott stood firm in his denials of all such allegations while before the AIB, and he was equally adamant that there had been no shortcuts or compromises made in the quality of work on the aqueduct project to date.

On August 31, after hearing testimony from project attorney Mathews and a number of businessmen who had entered into various subcontracts with the city, as well as aqueduct workers and foremen, the AIB delivered its report. There was a bit of attendant controversy when the AIB requested funds to publish several thousand copies of the report for public distribution and the City Council refused, but that matter was temporarily forestalled when the AIB leaked significant portions to the press.

In its report, the AIB concluded that the resources of the Los Angeles River had been significantly underestimated—in direct contravention of Mulholland's statistics, the AIB declared that waters drawn from the original watershed could support as many as 1,000,000 citizens (at the time there were about 350,000 in the city and 630,000 in the county). More alarming was the conclusion that the waters of the Owens River were contaminated by manure, sewage, and fertilizers, not to mention "drowning animals of various sorts," and were categorically "unfit for drinking purposes," making the entire "11,000 or 12,000" inches of water that might be obtained via the aqueduct useful only for irrigation purposes. These findings led inevitably to the further conclusion that, "the owners of the *Times* and the *Express* [owned by businessman and land speculator E. T. Earl], and wealthy associates . . . were interested in San Fernando Valley and other lands which would naturally be benefitted by bringing the Owens Valley Water to the head of San Fernando Valley."

As to the charges of shoddy construction practices, the AIB declared that there was "a general lack of supervision," and that "costly experiments" were made that resulted in "immense loss" to the city, including the use of tufa cement, considered substandard by consultants employed by the AIB. In addition, the once-touted caterpillar traction engines had finally proved to be unreliable in the desert climate, subject to regular breakdown and expensive repair. An exasperated Mulholland had finally ordered the machines be replaced with far less finicky mules, but in the AIB's mind, purchase of the engines constituted an unfortunate, indefensible experiment to begin with.

However, the body had heard nothing that indicated malfeasance on the part of any public official connected with the project: "no direct evidence of graft has been developed," the report concluded. However, the board did hedge somewhat by adding that "the Aqueduct system affords opportunities for graft, and that if this Board had the necessary time to develop all facts along lines suggested by individuals, a knowledge of human nature indicates that men would have been found who had succumbed to temptation." The codicil was a veiled rebuke of Mulholland and a reflection of the traction that the conspiracy theory had gained.

Coverage by the Socialist-leaning *Los Angeles Record* featured headlines such as "AQUEDUCT WATER IS POISON" and "MILLIONAIRES PROFIT—CITY PAYS," while the *Times* led with

"ENGINEER PROTEST AGAINST FAKED UP 'REPORTS'" and "INVESTIGATION WHIRLIGIGGLE—PROBE GOES AROUND IN VACUUM CIRCLE." Mulholland had little to say about the report publicly, though he did complain about what he said were "willfully garbled and distorted reports of the sessions which had found their way into print."

Meanwhile, the report was in the hands of the City Council, where it languished until January 21, 1913, when the council finally approved the funds necessary to print 10,000 copies in a newspaper-style format. Though no action was ever taken as a result of the AIB's findings, the process, which took more than a year to complete, did lay the groundwork for a bifurcated view of the aqueduct project that endures to this day. In certain quarters, the AIB report—lacking any evidence of collusion or graft and ignoring all known water science of the day—would be held up as proof that the undertaking was a flawed-from-the-outset scheme hatched by oligarchs to make millions. The opposing view, that the Los Angeles Aqueduct was a visionary enterprise engineered by a public servant the likes of which the world had never seen before, may have simply been too inspiring for cynics to tolerate.

IF YOU DIG IT, THEY WILL COME

ONE OF THE MORE LAMENTABLE EXCISIONS MADE from Catherine Mulholland's original manuscript is that of an entire chapter entitled "Thoughts on the Aqueduct Controversy," in which the author ponders the immensity of the discord that lingers about the subject. "The sheer magnitude and excellence of the engineering feat has often been diminished (even ignored or dismissed) alongside the human conflicts it generated," she writes, elsewhere theorizing that "suspicion remains that there must have been villainy, for the equivocal nature of the case dissatisfies those who demand that the event be a simplistic story of good vs. evil." Better, she counsels, to adopt a mythic stance toward such material, quoting Joseph Campbell to explain. "Myth looks with a godlike gaze at the hardness of life itself," she points out. It is a lens that, in Campbell's words, "makes the tragic attitude seem somewhat hysterical, and the merely moral judgment shortsighted."

Her grandfather "was never tortured by the ambiguities of what he had to do," Ms. Mulholland writes. "He told his sons in their early trips to Owens Valley before the Aqueduct was completed and when it was still a hunter's and fisherman's paradise (my father never forgetting the trout that fairly leaped to the line) that you couldn't let sentiment stand in the way of progress."

In sum, she argues, "while he oversaw the gigantic undertaking of the Aqueduct, Mulholland may have sometimes failed in 'displays of altruism' but he also avoided the pitfalls this project offered in the way of corruption, cheating, favoritism, and failure. The man with a clear head is not always lovable."

However dedicated he may have been to the building of the aqueduct, the various inquisitions of investigatory boards were not, by any means, the only matters Mulholland had to contend with during the long year of 1912. His wife, Lillie, had been diagnosed with uterine cancer, and given that there were now five children in the household—two not yet ten—domestic concerns were also taking their toll. Though much of the difficult work on the aqueduct was completed, key matters remained, and the work had begun to wear on him. As he told one reporter, "strain and responsibility have shattered my health. Now, under my doctor's advice, I am trying to forget everything connected with the aqueduct. My work is nearly completed, and then I shall take a long rest."

In May, however, he was described as cheerful when he announced to reporters, "This morning only twenty-five and one-half miles of the Los Angeles Aqueduct remain to be built," and told them that the route of the main distribution line for the system was determined—southward from the site of the reservoirs underway in the San Fernando Valley near the Newhall Pass and through a tunnel in the Santa Monica Mountains at Franklin Canyon to serve what is now Hollywood and the Inglewood area. He could possibly complete work on this part of the delivery system by the time the

Owens River water arrived, but appropriations would have to be made soon, he pointed out.

Meanwhile, there were still more than 2,000 men busy along the entire line, doing perilous work on the siphons and tunnels in the Saugus, Jawbone, and Grapevine Divisions, where they had been lucky to escape without serious mishaps so far. It was a rather surprising run of luck, in fact, given the nature of the work, and the terrain. Then again, good fortune is only known by what opposes it.

On the morning of June 16, in the Drinkwater Tunnel in the San Gabriel Mountains, about fifteen miles north of Saugus, there were about twenty men at work in the southern half of the 5,600-foot shaft known officially as Tunnel No. 9. A little before 8:00 A.M., foreman Lewis Gray led a blasting crew of three others to a point about 2,200 feet inside the mountain. About 400 feet of rock separated Gray's crew from the men working down from the opposite end, and they were scheduled to break through on July 1, assuming nothing unforeseen took place.

At the blasting site, Gray went over the task at hand with the crew. They were to place fifty pounds of dynamite in holes that had already been prepared, apply the priming powder and caps to the charges, and run the fuses in readiness for the blast itself. It was nothing the men—shift boss Norman Stoble, tool nipper Thomas O'Donnell, and rock train man Edward Garside—hadn't done any number of times, but still Gray left them with an admonition: "Be careful."

Gray couldn't be sure what happened after he rounded a curve that hid the men from his sight. There was electric light in the tunnel, and even if the lines didn't run right up to the spot where the trio was working, there was plenty of light there to work by. Still, Gray knew, miners were creatures of habit. Though it was absolutely forbidden, the men liked to carry their candles and often used them to supplement the electric lamps. Or maybe it was just a spark thrown as someone crimped a cap on a charge.

All Gray knew for sure was that less than five minutes after he'd left the group, and as he was still making his way toward the mouth of the tunnel, he heard a tremendous roar. In the next instant he was flying through darkness, then slammed into a wall of rock. Gray was bleeding, stunned by the force of the blast and barely able to breathe the air that was filled with dust and fumes. Still, he forced himself to crawl toward what he hoped was light. To stay where he was would mean death.

The men he'd left behind were not as fortunate. Stoble and O'Donnell had been working together a few feet away from Garside, who himself was only inches away from the charges at the head of the tunnel when the blast went off. The bodies of Stoble and O'Donnell were eventually found, buried beneath the forty carloads of granite that came down in the cave-in caused by the explosion. Nothing was left of Garside.

Most of the other members of the shift, working closer to the mouth of the tunnel, felt the blast and had hardly turned to run when the lights went out and the roof came down, jamming the passageway—twelve feet wide and thirteen feet high—before them. As they tore at the pile of fallen rock, foreman Gray emerged bloody and battered from the cloud of dust and fumes at their heels. He told them what had happened, ignoring his injuries to join in the task of dislodging the jumble until finally—heads splitting from the effects of the fumes—they could see light and could crawl and stagger forward and were out of the tunnel and able to breathe at last.

The survivors were soon joined by a hundred other members of the work camp hurrying back inside the tunnel to search for survivors. Deep inside they found half a dozen men who had turned the wrong way in the darkness and confusion. Instead of escaping the fumes, they had run deeper into them, and, by the time they realized their mistake, were too weakened to go on.

In time, the bodies of Stoble and O'Donnell were uncovered from the muck. Although O'Donnell's clothing had been blown from his body by the explosion's force, a man found his watch nearby, the hands stilled. In that manner, the time of the blast was officially determined. Lewis Gray was the son of John Gray, chief engineer of the division where the blast had taken place. O'Donnell had come up from San Bernardino, grateful to have found work. As for Garside, not only had his corporeal being vanished, no details of a family or a past could be ascertained. It was the worst accident to strike the project in its time.

There was one footnote to the tragedy worth mentioning. In the aftermath, as bodies were being carried off and the injured tended to, one of the rescuers heard a strange sound issuing from the mouth of the tunnel. He walked closer, then turned to a companion. "It's Maude," he said, and soon members of the crew were dashing back inside the tunnel behind him.

Maude was the miners' pet mule, part of the original four-legged crew that pulled the muck-laden cars out of the tunnel before the arrival of electricity rendered her an anachronism. The miners, though, had made a favorite of the even-tempered and tireless Maude. They allowed the mule, who often had to be forced out of the tunnel at the end of a shift, to keep on working. She had been back inside the shaft that day, patiently awaiting the resumption of her duties when the blast came and knocked her unconscious. In the darkness and confusion, rescuers had forgotten about Maude until they heard her braying.

Finally, they found her deep inside the passage, bruised and filthy, but otherwise intact. When the men tried to lead her out of the tunnel, however, Maude splayed her legs in the fashion of her kind and would not budge. The men had to bring food and water down to her until the car line could be cleared and Maude herself could be hauled back up top.

As a result of the mishap, which a coroner's report would declare an accident, Mulholland immediately declared the institution of stricter safety measures regarding blasting operations. At a Chamber of Commerce dinner on June 18, he assured an audience of 2,500 that ongoing work was of the highest quality, that Owens River water was "beyond question," and that the quality of cement used to line the aqueduct was unquestionable. As to the report of cracks in the line, Mulholland responded that it was "an absolute impossibility to build mile after mile of concrete work in burning heat and freezing cold" without expecting cracks. "Yes, there are cracks," he said, "but all the water which seeps through those cracks will not give the smallest desert dicky bird a respectable bath."

There had been water flowing in forty-five miles of concrete-lined conduit for as long as two and a half years, he said, and there was no appreciable loss of water. As for the failure of the caterpillar engines, he would have to admit it was a mistake. But now that the water department's 1,200 "hayburners" (including Maude, one presumes) were back on the job, they were making up for lost time.

Despite the loss of another laborer in a Saugus Division collapse scarcely a week later, work continued steadily. By July 7, Mulholland told a reporter for the *Los Angeles Times* that only 18.99 miles of construction remained, though dredging operations had slowed in the Alabama Hills when operators struck "an unexpected pocket of gravel."

On August 10, J. B. Lippincott told reporters that the last tunnel on the project had finally been "holed through" at a spot about

50 miles north of Mojave, in the Grapevine Division, completing a total of 42.69 miles of tunnel work. Allowing for the Alabama Hills delay, the project was now expected to be completed by April of the following year. There were now one hundred miles of continuous line completed, and forty-eight miles of that had been filled with water and tested, Lippincott said; there remained only three steel siphons in the Saugus Division (the Soledad Canyon Siphon was the largest of them) to complete. The department had also established a storage and warehouse facility at the intersection of Slauson and Compton Avenues where it was already engaged in selling off surplus property, he said.

In an exhaustive rundown of cost savings and purchasing practices, Mulholland's assistant had earlier pointed out that virtually no detail had been overlooked in trying to keep operations under budget. "Moca and Java coffee which costs the ordinary mortal about 30 cents a pound is sold to the aqueduct for 18 cents," he said proudly.

In light of the steady progress, discussion had already begun regarding the most appropriate way to celebrate the arrival of the water, with some calling for a statue of Mulholland to top a memorial fountain in Exposition Park, just south of the USC campus. The fountain would fittingly be dedicated as Owens River waters first shot up from its jets, the culmination of a month's worth of activities and exhibitions that might cost as much as $200,000. Though this particular fountain would never be built, the very notion gives an idea of the excitement that Mulholland had engendered in the community.

But there was also discussion of a more urgent nature taking place in anticipation of the water's arrival, and that concerned the disposition of the excess water that the aqueduct would carry. Groups from Pasadena, Santa Monica, Glendale, San Pedro, Long Beach, Glendora,

Covina, and San Dimas—not to mention the San Fernando Valley—were petitioning to share in the new supply, though Mulholland remained opposed to the city's building supply systems anywhere outside its limits. The water was city property in his eyes, and though Los Angeles was within its rights to sell to other municipalities, it was by no means obligated to bear the cost of building some other town's water works. If a community wanted to be part of the city water system, the solution to Mulholland was simple: become part of the city, pay city taxes, and enjoy city services.

Obviously, many of the petitioners were opposed to annexation on one ground or another, and few had the wherewithal or public support for the cost of constructing elaborate distribution systems. Also in opposition to Mulholland was new water commissioner S. C. Graham, who took a more entrepreneurial approach. Graham wanted to build what became known as the "high line," a supply system that would run to the eastern San Gabriel Valley. In addition, Graham proposed that the city would recover its costs simply by charging higher rates for the water delivered. Additionally, if the city ever did find that it needed the water for use within its own boundaries, it could raise the rates charged to the outlying districts along the high line beyond their ability to pay, lending an unintended irony to his project's nickname.

Graham's plan may have been elegant in design, but in Mulholland's view, it was Machiavellian in intent. The water was worth the same to everyone, Mulholland believed, but he was far more interested in seeing the city grow than he was in getting into the water sales business. Certainly it did not seem fair to him to entice a community into a contract that could be revoked long after that community had become dependent upon a water supply.

To counter the Chief's opposition, Graham crafted a compromise proposal that required outlying districts to pay for their own distribution systems while agreeing to a fifteen-year contract to buy water from the city for $10 an acre-foot. Graham touted the plan as a sure money maker that would also require the dastardly interests in the San Fernando Valley to pay through the nose for any Owens Valley water they received, but the city's charter would have to be amended to allow for such sales, so a special election was called for November. Once again, Mulholland found his views spurned by voters, who, on November 5, approved the Graham Proposal by a margin of 2 to 1.

Negotiations began almost immediately between the city and towns along Graham's proposed "high line." The City of Pasadena proposed to buy one-fortieth of the total aqueduct water supply—or 500 inches of water—enough to ensure its supply for at least five years. Left up in the air was the question of whether Pasadena would be asked to pay either a proportionate share of the cost of the aqueduct as a part of the deal or simply to contribute a proportionate share of the cost of the high line itself.

All such euphoric talk was soon tempered by the fact that the city's attorneys determined that the high line could not be built with funds from the outlying districts but instead would have to be paid for by a bond issue, which would require yet another election, set for February 11, 1913. Until voters approved those bonds, all talk of a high line would remain speculation. Mulholland had not actively campaigned against the Graham Proposal in advance of the November election, but this time would be different. In a letter of January 25, 1913, to the chairman of the Public Service Commission and widely reprinted, Mulholland denounced Graham's plan for its "rapacious audacity," and also declared that he was as firmly opposed to the building of any line to service the northerly slope and west end of the San Fernando Valley as he was to building the high line into the San Gabriel Valley. "The city should not pay for either of them," he said.

He envisioned a total serviceable area bounded roughly by the northern slope of the San Fernando Valley, the westward intersection of the Coastal Range and the Pacific, the eastern boundary of Pasadena, and a wavering line drawn southward from there to San Pedro, comprising a total of about 195,000 acres. To think of extending water service beyond those bounds was completely impractical, he said, and Graham's ill-defined high line could cost in excess of a quarter of the amount it took to build the entire aqueduct. And even if the board could decide to withdraw water service from the high line one day, the city would be stuck with an $8 or $9 million "dead pipe" for which it had no use.

It seemed close to irrationality in Mulholland's eyes for voters to be stampeded into making a commitment to a project as open ended and ill defined as the Graham Proposal, in which not so much as a firm termination point had been established. Mulholland warned that the proposal could give control of "about one-half of the total flow of the aqueduct to a region that has never sought, but always resisted, any talk of annexation to the city of Los Angeles, and from which not a single drop of return water can ever be recovered to the future use of this municipality."

In taking this stance, Mulholland was pitting himself not against amorphous outside forces but against his very employers, the Public Service Commission and the Water Commissioners. By insisting that water should be sold to all at a fair price, that once water service was given it should never be taken away, and that the citizens of Los Angeles were the owners of the aqueduct waters, he presented himself squarely as a defender of the public interest. But in asking voters to make a choice between principles and the chance to make a quick return on their investment promised by Graham, he was taking a big chance.

LAST MILE

WHILE POLITICS WERE PLAYING OUT, THERE WAS still work to be completed up and down the line, and that work remained dangerous and demanding. On the evening of January 19, 1913, three men, identified by the *Inyo Independent* as "two Mexicans and one American" ventured inside a section of steel pipe on the Jawbone Division, carrying buckets of asphalt oil paint used to coat the inner as well as the outer surfaces of the pipes. It was not particularly pleasant work, owing to the close quarters and the noxious odor of the material they were working with, but laborers on such a project were rarely deterred by such considerations. They were being paid well, there were days off in Mojave to look forward to, and there were plenty of men waiting to slap that tar if they chose not to.

It is not unlikely that the trio assigned to paint the inside of the siphon had heard of the story recounted recently by the *Los Angeles Times*, the tale of how a man named Ratich, "a Slovanian," who, lacking the price of rail fare, had attempted to walk from Los Angeles, by way of Saugus, up to the aqueduct camp at Mojave where he hoped to find work. Ratich lost his bearings and ended up wandering the desert near China Lake, where someone riding a passenger train noticed him. At the next stop, alerts were wired back up the line and a search party set out around 2:00 P.M., when the heat was at its peak. By 6:00 P.M., a section foreman named O'Malley patrolling the dusty back roads in his car was ready to give up and go in when he spotted a lone figure in the wastelands nearby.

O'Malley stopped and approached the man, who stared back warily. He eyes were bulging, his tongue was dangling, and blood dripped from the tips of his fingers. O'Malley had seen it before: the man had torn off his nails digging desperately in the sands for water. When O'Malley told Ratich who he was and that he had come to help, Ratich responded by charging him in a fury. After some moments of struggle in the dusty desert floor, O'Malley was able to get himself atop the weakened man and half-dragged, half-carried him through the desert to the nearby rail station. It would take two more men to subdue Ratich when he spotted the water cooler sweating beneath the eaves.

Given stories such as these, it seems unlikely that any of the men who clambered inside the Jawbone Siphon on the evening of January 19, 1913, ever expected that life would be easy, though none of them probably assumed it would necessarily be short. In any case, theirs became so when one of them lit a candle or a cigarette or somehow struck a spark shortly after the painting had begun. There would have been scarcely an instant between the swipe of a match across denim, perhaps time for someone's glance of shock or surprise or curse, and then the white light of explosion that culminated in a headline: "AQUEDUCT GAS KILLS THREE—NAMES AND OTHER DETAILS LACKING." As one report laconically concluded, "The fire warmed some of the plates of the siphon slightly but not enough to do any harm."

In all, there were forty-three men killed in accidents during the course of the aqueduct's construction, according to Dr. Taylor's figures. One accident resulting in "permanent injury" had taken place, and there were a total of 1,282 incidents where treatment was recorded. Two men were reported to have died from disease, though the medical statistics are known to be incomplete. As to the number who might have succumbed to forces unknown on the dark streets of Mojave where a rum-running sheriff kept the tally, who can say?

Typical among Taylor's cases was that of the cement plant superintendent who happened to be passing one of the machines that ground the tufa rock into powder that could be mixed with cement. The grinders were driven by wide belts, one of which snapped just as the man happened past. The broken end whipped off its spool and struck the superintendent in the back, driving him into a nearby steel framework where a pipe broke his skull.

The badly injured man was placed on a stretcher and taken to the nearby Southern Pacific station where an emergency train was sent from Mojave to transport him to the California Hospital in Los Angeles. There Taylor and another surgeon awaited. "We found that the whole dome of his forehead had been caved in and quite a number of pieces of the frontal bone . . . had been pushed into the cranial cavity," Taylor recalls. "His brain had been very definitely damaged . . . portions of it spilling down onto his face."

The two doctors "removed some pieces of bone, replaced what we could, sewed him up, and put him to bed." Though they were uncertain as to his prospects for survival and expected that he would be seriously compromised at the very least, the man "never had a rise in temperature or untoward symptoms, not even a headache," Taylor says. In fact, he regained consciousness quickly and was back at work within a few weeks. Years later, Taylor saw the man, who had returned to Los Angeles after a sojourn running a cement mill for a New York firm in Argentina. "I asked him if he had any bad effects from his brain injury," Taylor recalls.

"None whatsoever," the man responded. "Except for the one thing."

And what was that? Taylor wanted to know.

"I did lose my libido, and my wife left me," the man said. Whether the latter was a bad thing remained unsaid.

THROUGHOUT THE FIRST QUARTER of 1913, Mulholland was overseeing work on the aqueduct, on the construction of the San Fernando Valley Reservoir, and on the Franklin Canyon pipeline and reservoir. His opposition to the Graham Proposal prompted the mayor to ask the City Council to reschedule the bond-issue election from February to April 15 so that the matter could be "more carefully discussed."

With a mayoral election coming up in May, some Progressive Party members began to press Mulholland to consider running for the post. While his bon mot in response to this suggestion—"I would sooner give birth to a porcupine backward"—has been given wide currency, Catherine Mulholland passes along a quote from Mulholland's correspondence of the period that gives something of a deeper insight into the man. In a letter to longtime Los Angeles politico Charles Willard written on January 24, Mulholland said that those who were urging him to run should "recognize my temperamental unfitness for the position of Mayor." Perhaps he had managed to hide it, he continued, "but it is nevertheless a fact that in the execution of my work I have tendencies that are absolutely autocratic and at times unreasonably domineering."

Even if he had been successful at concealing this trait, Mulholland continued, "This I could not do, of course, in the position of Mayor of the City where events move so rapidly that my impetuosity under necessarily quick action would reveal my weakness." He was proud of the fact that he had always been able to win the loyalty and devotion of coworkers, Mulholland concluded, but he was sure that "in the discharge of the multifarious duties of Mayor I would utterly fail in this particular."

Soon after he delivered this piece of self-analysis, a *Times* story appeared (February 9, 1913) to tout the work being done on the San Fernando Reservoir in the valley about a mile and a half south of the terminus of the aqueduct near the Newhall Pass. The dam, 740 feet wide at its base and 140 feet high, would be the second-largest in the world at the time, creating a lake a mile and a half long and a mile wide.

Fill for the dam was being put in place using the hydraulic methods that Mulholland had perfected. "First the dirt is loosened by charges of powder, a carload of which is used every month and the 'back fire' of water is then permitted to run into the disintegrated particles of sandy loam," the story explained. Then water piped down from Soledad Canyon was used to jet the soupy mix into place: "When the hydraulic streams are turned on from in front there is not much left for that particular hill to do but move out," the reporter said.

In all, the process allowed the dam to be built for about one-sixth of what it would have cost using the old steam-shovel-and-wagon, old-women-carrying-dirt-in-their-aprons method, project superintendent Stanley Dunham said. Though work on the aqueduct itself was nearing completion, as was the dam, Mulholland told reporters that it would take another six months to complete the pipeline that would connect the dam through the Franklin Canyon to the city's water system.

Excitement continued to mount in Los Angeles as word traveled down from the Owens Valley. On February 13, "While the snow-crowned peaks looked on," and "a woman in furs" smashed a bottle of champagne atop the concrete gates, "eight men turned iron wheels that opened the flood gates and the cold crystal waters of the Owens River rushed through the intake into the upper sixty-eight-mile division of the Los Angeles Aqueduct." Among the eight, predictably, were Mulholland and J. B. Lippincott, with Harvey Van Norman joining them as well. In fact, the fur-clad woman was Bessie Van Norman, the engineer's wife, the same person whose thwarted arrival in the valley years before had nearly caused Van Norman's resignation.

The sixty-mile Owens Valley portion of the aqueduct beginning at the diversion gates above Independence and ending at the Haiwee Reservoir below Owens Lake was now complete. Beyond the four gates, each 8 feet high and 8 feet wide, the water would ride about 20 miles in an unlined canal 62 feet wide and 10 feet deep, to allow the flow to absorb the considerable amount of groundwater percolating in that area. At the Alabama Hills, the aqueduct became a 40-mile-long concrete-lined canal 34 feet wide and nearly 13 feet deep that hugged close to the mountains, overlooking its former destination at Lake Owens, until it reached the Haiwee Reservoir, 3,760 feet above sea level.

The dam at Haiwee was built across the canyon where the Owens River had run in ancient times and had walls at both its lower and upper ends, containing a reservoir more than seven miles long and anywhere from a quarter to a half mile wide. The artificial lake would hold 21 billion gallons of water, enough to supply the city for sixteen months at its then daily consumption of 40 million gallons a day. With the gates open at full-bore, it would take nearly two months for the reservoir to fill.

"It was a great sight to see the water rush through," Lippincott told reporters at the ceremony. The party intended to follow the first gush of water all the way to Haiwee, he said, though they would be staying overnight at Lone Pine in the process, for it would take the waters, flowing at about two miles per hour, a day or so to get to the new reservoir. From that point, all that remained to connect that vast new lake with Los Angeles, Mulholland said, was about a half mile of shovel work, three miles of ditch lining, and the completion of one 2,000-foot tunnel in San Francisquito Canyon, all of which he thought could be finished in the following six weeks. As has been noted, however, aqueduct affairs would never again be free of political intrigue that complicated matters. Though Mulholland had bowed out of consideration, the mayoral race hinged in large part upon aqueduct issues, prompted in part by the final publication of the Aqueduct Investigation Board's report in January. In addition, controversy continued in advance of the high-line bond issue, which was finally rescheduled for April.

The *Times* reprinted Mulholland's scathing letter to his bosses as part of a two-page spread on April 13, just two days before the bond election. The results were a vindication of the Chief's passionate statements. On Question #4, as to whether to issue $2.5 million to begin construction of the "high line," some 15,000 answered in the affirmative; more than 32,000 said no. As to the $1.5 million in bonds needed to complete work on the connection of aqueduct water, the vote was 46,792 for, 4,798 against. If the Chief had indeed been wondering if the people had stopped listening to him, the results would have erased such doubts.

As to the mayoral election the following month, Job Harriman ran again for the post, but the effects of the McNamara case had never left him; he finished third behind two Independent candidates, neither of whom was able to reach a majority. Henry Rose won the ensuing runoff, but only when a considerable number of Socialists swung his way in exchange for a promise to "clean up" operations in the aqueduct department. The day after his election in June, Rose announced that he had learned of greater waste on the project than he could ever have imagined and declared that he would make a personal inspection of the line later in June. Following that trip, however, Rose returned to the city to tell a reporter for the *Los Angeles Examiner* on September 16 that he had seen no evidence of the rumored pollution, poor construction methods, or inefficiency. All such criticism, "so far as I am able to determine," Rose said, "is captious."

Despite Rose's report, there had in fact been one miscalculation on a difficult portion of the project, however. Since March, Mulholland had been concerned with the huge 2,800-foot siphon at Sand Canyon, which with its drop of 442 feet below grade made it the second deepest after the Jawbone. Unlike the other siphons, which had been constructed of concrete or steel pipe, the Sand Canyon Siphon, about 100 miles below the intake point and 150 miles north of Los Angeles, consisted of two tunnels bored precipitously down through the slopes of opposing rock walls (the bores were grouted with concrete) with a steel pipe running across the canyon floor to connect them. The bore on the north side was about 640 feet long and 9 feet in diameter and descended the canyon at a 45-degree angle. Water ran in a pipe across the canyon floor for about 1,500 feet, then entered the opposing 45-degree bore for a climb of 632 feet back up to grade. Because the cliffs were solid granite, so hard that miners needed to up the charge to blast through by 25 percent, it was assumed that the tunnels would be as strong as, if not stronger than, pipes of inch-thick steel.

However, when engineers began to test the line on the north side of the canyon, ominous seepages appeared in the sheer rock wall there. When the tests were repeated on the south side of the canyon, similar seepages developed. On Monday, May 19, Mulholland returned and gave orders to send waters through the siphon at full force.

As everyone waited apprehensively, the torrent gushed down the north side of the canyon through the bore, then hurtled on through the steel pipe on the canyon floor and began its rush up the tunneled rock on the opposite side. There then came a thunderous roar as—impossibly—the solid rock wall before them blew apart, sending thousands of tons of boulders and fractured rock to the valley floor below. In all, nearly half a mile of tunnel—bored through what had appeared to be solid rock—was ruptured almost instantly. "The incline tunnels failed spectacularly," as Mulholland put it, and "the side of the mountain was lifted bodily and shattered into a mass of debris."

The dramatic setback—which thankfully cost no lives—did not take Mulholland completely by surprise. He told reporters, "That is the only point in the whole system that I have ever had the slightest doubt about," he said. That's why he had come up, "for the very purpose of giving it a good kick and making sure." As for the rupture, he said, it was better to know ahead of time rather than when the water was flowing to the city. Though the north side of the rock siphon had held, they would take no chances. Both rock tunnels would be abandoned and replaced with steel pipe, Mulholland said, at a cost of about $28,000. It would also mean a delay of at least three months, which would postpone the arrival of the water to the San Fernando Reservoir until fall.

Though a program describing the planned festivities had already been drafted, it was scarcely any concern of Mulholland's. Until his work was done, celebrations would simply have to wait.

CASCADE

It was late September 1913 before the new steel siphon was completed at Sand Canyon, a task that constituted the last remaining item of significance on Mulholland's checklist. At noon on Thursday, September 25, and following a visual inspection of the new pipe, Mulholland phoned from Sand Canyon up to Haiwee, instructing his men to open the gates at the reservoir. It took seven hours for the water to travel the twenty-five miles or so to Sand Canyon, but when it arrived, the big pipe held. By the following morning Owens Valley waters—about 3,000 inches, or an eighth of capacity—had passed through the siphons of the Jawbone Division, seventy-four miles farther along. By Friday night, traveling at about three and a half miles per hour, the water reached Mojave.

Meantime, Mulholland drove from Sand Canyon up to Haiwee and took a spin on the recently formed "Lake Mulholland" there in the department's new boat, then traveled back down the line, checking tunnels, siphons, and conduits for problems. The pipes had held, and Mulholland observed wryly that none of the tufa-added concrete had disintegrated when the water hit it. At least one other thing had been proved, he told a reporter: "The aqueduct does run downhill from the Owens Valley."

The correspondent noted that he had sampled the water at Mulholland's invitation, both at the Haiwee Reservoir and out of a pipe in the Jawbone Division. He found it "good," the reporter cabled. "No symptoms of alkali poison apparent. Hair not falling out. Owens River will be flowing down San Francisquito Canyon Saturday night."

It was something of an overstatement on the reporter's part, for no water would be permitted to flow beyond the Fairmont Reservoir until the impoundment there was full, a process that took until October 1. From that reservoir, the waters were turned through the five-mile-long, ten-by-twelve-foot Elizabeth Tunnel, then shot out of the South Portal and down a fifty-foot fall to the bed of San Francisquito Creek, which it would use until the last of the conduits and power generating facilities in the canyon were completed. The flow was finally halted at Dry Canyon Dam about half way along the thirty-eight-mile stretch to the Cascade above the San Fernando Reservoir. There the waters would be held until released for the opening ceremonies, now set for November 5.

Mulholland was already heartened by the fact that Mayor Rose had swung over to full support of the project, ousting water commissioner Graham from his position soon after he had taken office. The Public Service Commission had responded to the failure of the Graham bond issue with an essential endorsement of Mulholland's long-held position—only those areas likely to be annexed by the city and capable of being economically served would be eligible to purchase excess water.

In addition, Mulholland was also pleased by an unexpected discovery in the wake of the opening of the gates. The aqueduct was running at the rate of 25,000 inches, a quarter more than he had projected. For one thing, there was far less friction along the line than he had allowed for, and there was virtually no leakage to be found. If that rate of flow held, it would mean a considerable financial advantage to the city, for it meant that much more excess water could be sold.

On October 30, details of the celebration planned to mark the aqueduct's completion were published in Los Angeles papers, including elaborate festivities at the Cascade on Wednesday, November 5. "The big job is finished," Mulholland told reporters. "Nothing remains now but to shoot off a few firecrackers, turn on the water and tackle the next big job."

It was perfect weather for the celebration on November 5, with the nighttime low rising from the mid-fifties to a pleasant seventy-two by midday. Somewhere between 30,000 and 40,000 citizens turned out at the Cascade below the Newhall Pass to celebrate the arrival of water that had been promised to them for eight years—the building of a delivery system that had taken six of those years was now complete.

For Mulholland, it was a day of mixed emotions. He was proud, of course, but he was also concerned about the health of his wife, Lillie, who had just undergone an operation for the uterine cancer she battled. His daughters, Rose and Lucile, sat with him on the reviewing stand built in the arid hills near the spillway gate and at one point during the ceremonies, word was passed to them that Lillie had, in the opinion of her doctors, pulled through the crisis.

After various preliminaries, including a band concert, a military salute to the arriving dignitaries, and several introductory speeches by politicians and by his assistant J. B. Lippincott, the fifty-eight-year-old Mulholland rose to speak. His remarks were typical in their brevity. "The aqueduct is complete and it is good," he said. "No one knows better than I how much we needed the water. We have the fertile lands and the climate. Only water was needed . . . and now we have it."

He was also characteristically generous in the kudos he gave Fred Eaton, who, still rankled by Mulholland's opposition to his various attempts to sell land to the city, did not attend the ceremony. To the former mayor, Mulholland gave unequivocally "the honor of conceiving the plan of the aqueduct and of fostering it when it most needed assistance."

He concluded by saying, "On this crude platform is an altar to consummate the delivery of this valuable water supply and dedicate to you and posterity forever a magnificent body of water." With that, Mulholland pulled on a lanyard that unfurled a big American flag above the platform, a signal that called forth a volley of cannon fire and aerial explosions. Higher up the hillside, General Adna Chaffee, Harvey Van Norman, and others began to turn the wheels that would unleash the water.

It took two or three minutes, one reporter said, before anything happened, but "then the water came" in a flood that filled the concrete stair-steps of the Cascade more than two feet deep. The crowd, as reporters noted, "went wild with delight." It took everything in the celebration chairman's power to restore order so that Mulholland could go back to the podium to mark the formal conveyance of the aqueduct to the City of Los Angeles. Mulholland was not one to disappoint. He gazed out at the crowd and gestured toward the water. "There it is," he said, delivering what was certainly the line of the day, perhaps his most memorable line of all. "Take it!"

The celebration continued throughout the day and into the night. Mulholland and Lippincott were presented with silver loving cups by the Chamber of Commerce, and Mulholland, whose building project was touted as the equal of any ever undertaken, heard himself lauded as the "Goethals of the West," after the chief engineer of the Panama Canal.

At an evening banquet, the Southern California Association of Architects and Engineers gave Mulholland a parchment scroll of appreciation, emblazoned with a copy of the Roman aqueduct at Segovia, calling his achievement "stupendous," and praising Lippincott and Fred Eaton as well. In his own typically brief acceptance speech, Mulholland thanked the original water board for their courage in backing the project to begin with and attorney William Mathews for managing the complex legal and financial affairs, adding the quip about Mathews keeping him free to work and out of jail. As to those who criticized the project along the way, Mulholland claimed that he had never paid much attention to the naysayers, knowing that when the water finally arrived, all would be forgotten.

Kudos also came in from afar. C. A. Shaw, one of the three commissioners of New York City's 163-mile-long Catskill Aqueduct, budgeted at $170 million, lauded the foresight of Los Angeles, calling it "one of the most provident cities in the work of securing a water supply that will last for decades." Shaw pointed out that were the City of Los Angeles to have waited and been forced to condemn private property for the rights-of-way at some later point in time, the costs would have likely put the project forever out of reach.

In all, the aqueduct was a more than 233-mile-long system: about 24 miles of open, unlined canal below the intake, and another 37 miles of lined, open canal leading to the Haiwee Reservoir; once the final conduit and tunnel work was completed in San Francisquito Canyon, there would be about 98 miles of covered conduit, nearly 43 miles of tunnels (some of them large enough to have driven trucks through), 12 miles of siphons, another 9 miles of tunnels used in the power generating systems, reservoirs totaling some

8 miles in length, and various flumes, bypasses, and outlet pipes. There were as many as 3,900 men working together on the project, alongside the 1,305 "hayburners" who endeared themselves to the Chief far more than the traction engines did.

Mulholland had signed checks for as much as $575,000 in a single month (May 1913), but in the end, he had held expenditures within $100,000 of the original $24.5 million, an accomplishment modern-day project superintendents might only dream of. In fact, as Mulholland pointed out in a report to the Board of Public Works, when one figured the value of the 125,000 acres of Owens Valley land owned by the city, the value of the cement plant and surrounding lands ($550,000), the value of surplus equipment, the value of the three power-generating plants in the valley, and the 755 mules he had purchased rather than leased, he had actually *beaten* the original estimate by about $3 million.

Of the $24.6 million spent, just about half, or $12.5 million went to payroll, $8.15 million to materials and equipment, $2.25 million for freight, and just $1.7 million for lands and rights-of-way. Even presuming that the city kept its land holdings and power-generating plants in the Owens Valley, Mulholland estimated that the sale of the cement plant and surplus equipment would return $1 million to the city's coffers.

A reporter sent by the *Times* asked Mulholland if he was looking forward to a rest now that the great project was done. Mulholland thought about the question for a moment. "I took a vacation once, when the old board was giving me some trouble," he said. "I told my secretary, I'm going to San Francisco. I don't want a telephone call or a message. I'm going up there to rest and forget the aqueduct is on earth. I'll be back in two or three weeks."

It was a Monday when he left, Mulholland recalled. On Wednesday morning he walked back into his office. One of his assistants glanced up in surprise. "I thought you were going to San Francisco for a rest," the man said.

"I did," Mulholland answered.

But he couldn't have been there long enough for a meal, the man protested.

He wasn't, Mulholland allowed. "You see, I got up there, and I didn't know anybody, and I started thinking about all my friends down here, and about the job, and . . . well, my resolve broke." He shrugged. "I'll make the fight without the rest," he assured his assistant.

If the tale seems the emblem of a workaholic—indeed, then seventeen-year-old daughter Lucile described him as the man who came to dinner with the family "occasionally"—it was not a simple issue in Mulholland's case. In his eyes, he had been blessed with the opportunity to do work that he loved. If some men dreamed of being free of their work, he looked forward to doing his, just as he enjoyed the chance for a conversation on topics ranging from etymology to the roles of Sarah Bernhardt.

Questioned by the reporter as to his religious beliefs, Mulholland responded, "The golden rule."

As to his politics: "Conscience, progress, a chance for every man."

And as to his philosophy: "The inevitable is the inevitable. And work."

With his wife, Lillie, in the hospital, he rose each morning at six to call and check on her condition. By seven he was in the office, to dispense with mail and messages long before anyone else arrived. By eight or nine, he was in consultation with his supervisors, and from there the day would fly.

Even in his late fifties, he was described as powerful, with a broad chest and solid limbs, "a fighter's jaw, a thinker's head, eyes that are appraising, searching, thoughtful, sympathetic." A typical

Mulholland entrance was captured by the reporter hurrying along in his wake following a visit to the dam site at San Fernando.

"You have a banquet engagement at the City Club at six," his secretary called as Mulholland passed, trailed by several assistants needing just a moment of his time.

He would not be going to any banquet, Mulholland said. He was only there for a minute, then he would be going to the hospital to see his wife. "You can call that meeting off."

"And you don't have to go to Independence tonight, after all," the secretary added.

"I wasn't going anyway," Mulholland said.

The secretary seemed unfazed. The Chief should also be aware, the secretary said, that a certain engineer in a particular district had just fired an obviously familiar worker named Mulvaney.

Mulholland paused, lifting his battered hat to dust it off. "Drunk again?"

The secretary nodded.

"How long since the last time?"

"Eight months."

Mulholland nodded. "See if they have something for him down in Saugus," he said.

The secretary made a note. There had also been a call about someone named McGregor, she added, somewhat tentatively. "McGregor said the foreman had no business firing him, that he had a pull with you."

"Did the foreman fire him?" Mulholland asked.

"He did," the secretary answered.

"Then McGregor's fired," Mulholland said. "I'm going to the hospital."

IN THE SHADE OF ACCOMPLISHMENT

THE ARRIVAL OF THE AQUEDUCT WATER IN NOVEMBER 1913 might have been the most significant marker in a long and storied career, and in some ways it was for Mulholland, though his difficulties would scarcely end. In terms of the project itself, there was the hydroelectric work in San Francisquito Canyon to be completed, as well as the major conduit that would link the San Fernando Reservoir with the city's mains via the Franklin Canyon. In fact, the first Owens River water did not begin flowing through customers' taps until April 1915, nearly a year and a half after the ceremony at the Cascade.

The issue of the city's entry into the power business was also delayed until voters in May approved an allied $6.5 million bond issue to pay for power development, an issue vigorously opposed by long-time aqueduct booster Harrison Gray Otis and the *Los Angeles Times*. Meantime, the aqueduct itself had also taken a significant blow in February when heavy rains washed out supporting piers beneath a section of raised steel pipeline in the Antelope Valley. When the plates ruptured and the water escaped, an almost two-mile section of the ten-foot-diameter pipeline collapsed nearly flat, reviving critics' claims that the aqueduct would never function as intended.

Though some engineers on the project worried that the collapsed section would have to be replaced, at great expense and considerable delay, Mulholland remembered the relative elasticity of the steel pipe from the days of the test of the support system engineering. He simply ordered the break in the line welded shut, the supports replaced beneath the flattened pipe, and the water flow gradually turned back on. Bit by bit, the growing pressure in the line restored the pipe to its original shape, and within a month, the full flow of the water was returned. The total cost of what had been seen by some as a death knell was $3,000, and Mulholland's unheard-of fix revolutionized practices within the industry.

On May 12, as work on the sixty-eight-inch main from the San Fernando Reservoir southward was being completed, Mulholland boarded a train for San Francisco, where—variously proclaimed a "Genius," a "Super-Man," and "California's Greatest Man" in news headlines—he was to receive an honorary doctorate from the University of California at Berkeley. On the eve of his departure for the ceremonies, where George Washington Goethals, the chief engineer of the Panama Canal, and David Jordan, president of Stanford University, would also be honored, a reporter described Mulholland as "a most unconscious recipient of the plaudits," saying that he had attended that day's meeting of the Public Service Commission, "gave his attention to matters of trivial and momentous interest, smoked his usual cigar, and departed unruffled for the north."

Upon his return to Los Angeles, Mulholland found that two Socialists involved with the Aqueduct Investigation Board had filed a petition with a local judge for an injunction against further use of Owens Valley water on the grounds that it was grossly polluted and unfit for human consumption. Notwithstanding Mulholland's observation that the aqueduct water had been tested by experts on numerous occasions, not to mention the fact that thousands of men, including himself, had been drinking from it for years, a trial commenced, one that, despite the apparent flimsiness of any evidence, would drag on through various appeals until 1919. Mulholland, convinced that it was simply another veiled attack on the aqueduct by the private power interests, could do little but appear in court when called and patiently answer the vexatious questions of his cross-examiners.

At one point, an attorney asked if Mulholland could suggest where his clients could find creeks with polluted water in the Owens River Valley. "Oh, I could point you to far more polluted sources than that," Mulholland assured his questioner.

"Really?" the attorney asked, intrigued. "And just what would those be?"

Mulholland shrugged. "Well, you could draw some samples from a privy vault," he said. "Or you could roll a hog into the water with an apple in its mouth and take a photo of it."

Mulholland had never been one to suffer fools, but part of his impatience in such encounters could surely be attributed to his concerns with his wife's health. By April, Lillie's condition had worsened. Following a second operation, she lapsed into a coma, and on April 28, at the age of forty-seven, she finally succumbed to the cancer she had battled for years. Mulholland, who had always counted on his wife to manage matters on the home front, was both grief-stricken and baffled. In the end, his eldest daughter, Rose, born in 1891, took over her mother's role as the keeper of the house, one she would maintain for the remainder of her father's life.

Meanwhile, Mulholland threw himself into his work with a vengeance. Throughout the process of the aqueduct's construction, he had been irked by the waste involved in mixing concrete at the surface of tunnels, then carrying it inside in cars so that it could be shoveled out and used to line the walls. With significant tunnel work still going on in San Francisquito Canyon, a solution presented itself to him: Why not pump mixed concrete through pressurized hoses and spray it directly onto the tunnel walls? And soon that process, one that has become standard in the industry today, was at use in San Francisquito Canyon.

At another time, while preparing for a talk on the project before the annual meeting of the American Society of Civil Engineers, Mulholland was casting about for a way to dramatize the magnitude of the project for his audience. He hit upon the idea of replicating a topographical map of the aqueduct's route in three-dimensional form, using vulcanized pasteboard. The result—twenty-six feet long and two and a half feet wide, and exhibited at the 1915 World's Fair in San Francisco—was striking (it remains under glass at DWP offices) and inspired the process still widely employed in the mapmaking and modeling industry, including the impressive model of the region on display at the Eastern Sierra Interagency Visitor Center in Lone Pine.

The question of annexation of various communities into Los Angeles also remained on the list of Mulholland's concerns. By a vote of 681–25, the very few citizens of the dusty San Fernando Valley determined on March 20 to apply to become part of the City of Los Angeles, and on May 4, the citizens of Los Angeles approved, thus making it possible to sell water to irrigate some 100,000 acres, about half of the water the aqueduct would carry.

Development of municipal power had proceeded under Ezra Scattergood's direction and with Mulholland's support, but in fits and starts. An election on a bond issue that would allow completion of Power Plant #2 was defeated in 1917; completion would have to wait until after World War I ended, when voters finally approved the measure in a June 1919 election. In the end it would take until the mid-1930s before the city was firmly in control of its own power distribution.

On July 11, 1917, an earthquake shook the Owens Valley, one strong enough to cause a significant crack in a portion of the concrete conduit traversing the Alabama Hills. Then, four days later, there came two blowouts on the line, one in the Antelope Valley and another near Little Lake, south of Haiwee. Though there was speculation that the ruptures were caused by dynamite charges planted by opponents of municipal power, no proof of sabotage was ever discovered and no further such incidents occurred—not for a while, that is.

In the meantime, the glowing predictions of the city's growth proved to be shortsighted, if anything. Vast quantities of Valencia oranges that thrived in the newly irrigated soil of the San Fernando Valley were now being shipped back East and even to the United Kingdom by way of the Panama Canal. In 1919, nearly 50,000 carloads of citrus went out, along with walnuts, strawberries, wheat, pumpkins, celery, and olives, all of which could be grown throughout the year.

Sunkist began to market its revolutionary product called "orange juice" in 1916, and by 1918, nearly a quarter of the nation's supply of oil was being pumped from reserves beneath the city. Ford opened a Model T plant in Los Angeles in 1914, and Goodyear began making tires there in 1920, soon to be followed by Firestone, B. F. Goodrich, and others. The forerunners of aviation giants Lockheed, Douglas, and Boeing were established during the decade, and in 1915, a man by the name of D. W. Griffith shot a full-length feature film called *Birth of a Nation,* which engendered an industry that would become virtually synonymous with the place.

By 1920, the population of Los Angeles stood at more than 575,000, making it the nation's tenth-largest city, surpassing San Francisco (506,000) for the first time. Population in Los Angeles County had grown to more than 900,000 and would pass the 1 million mark in 1921. The area of the city had grown as well. In 1909, the port cities of San Pedro and Wilmington were annexed, and Hollywood was added in 1910, along with Cahuenga Township and a portion of Los Feliz. The annexation of San Fernando Valley lands in 1915 added 170 square miles, nearly tripling the city's size. In all, nearly 266 square miles of land were annexed to the city between 1915 and 1920. (The total area of Los Angeles would reach 469 square miles by 2004.)

They were heady times, but growth also brought with it an increased demand on city services, including those of the water department. In fact, by 1923, Mulholland was reassessing his earlier stated certainty that the waters of the Owens River would in fact assure Los Angeles of a sufficient water supply for the foreseeable future. By 1918, all of the available water carried by the aqueduct was being used, much of it going to irrigation, and an inevitable period of cyclical drought was about to begin.

On October 23, 1923, Mulholland asked the Board of Public Service to authorize a preliminary survey to determine the feasibility of a project to bring water to Los Angeles via an aqueduct from the Colorado River. Though key legislation regarding the plan to build Boulder (since renamed as "Hoover") Dam, Parker Dam, and the vast feeder system that would one day provide water via the Colorado River to Los Angeles, Long Beach, Orange County, and San Diego would not be approved until 1928, after his retirement from public service, it was Mulholland's vision that instigated another monumental undertaking.

Mulholland also sought more practical solutions closer to home. In his "Twenty-second Annual Report," dated June 30, 1923, Mulholland observed that the previous year had been one of the driest on record, and prudence dictated that steps should be taken: "No engineering corps having the important task of the City's water supply in mind would be justified in relaxing vigilance," he said. Since very little snow had fallen in the Sierra and the flow of the Owens River had diminished by nearly 40 percent, Mulholland proposed the purchase of more lands north of the intake point in the Owens Valley so that additional waters allocated to those lands could be diverted to augment the river's flow and groundwater beneath those lands could also be tapped.

By this time, property owners in the Owens Valley were well aware of the city's interest in acquiring additional lands and had formed an association of landowners designed to consolidate their common interests and drive up the price of any lands sold to the city. The most prominent of those associations was headed by two Owens Valley bankers, Wilfred and Mark Watterson, who held a virtual monopoly on credit and development in the area.

Complicating matters for Mulholland was Fred Eaton, who had held onto his ranch lands in the Long Valley (at the upper end of Owens Valley) in the belief that the city would one day be forced to build a long-discussed dam to capture the more reliable flow there. In order to create an impoundment of practical size, the city would need to buy his lands, and in Eaton's mind, given the growth of the city and the toll taken by the recent dry spell, the day when he would finally realize a return on his investment was fast approaching. He let it be known that the lands that he had acquired nearly twenty years ago could be had for the price of $1 million. To Mulholland, who had always considered Eaton well compensated for his efforts on behalf of the city, the price was outrageous. "I'll buy Long Valley three years after Fred Eaton is dead," Mulholland reportedly said to associates.

In refusing to deal with Eaton, Mulholland's high-mindedness put him between something of a rock and a hard place. Theoretically, the building of a 150-foot-high dam at Long Valley could have created a reservoir large enough to meet the city's needs *and* provide enough water to supply the irrigation needs of local farmers. Irrigated lands in the valley had increased from about 40,000 acres at the turn of the century to some 65,000 acres by 1910, and talks had been ongoing since that time in an attempt to reach a formal agreement guaranteeing the status quo, so long as valley residents would agree not to press for an expansion of those rights.

In 1921, the city finally reached an agreement with a citizens' group originally called the Owens Valley Defense Association, whereby it would construct a dam at Long Valley of sufficient size for the needs of both the city and the Owens Valley. Accordingly, in August 1923, work on the dam finally commenced, but it was to be only 100 feet high in order to avoid the need to purchase additional lands, including those held by Eaton. At that point, the Watterson group, of which Fred Eaton was a member, filed for an injunction to halt the work, though they let it be known that they would drop their opposition if the plans were changed to make it a 150-foot dam. For the time being, at least, stalemate prevailed.

It was ironic that Eaton, once vilified in the Owens Valley, had become one of its stalwart citizens. As early as March 1906, Eaton had told reporters of his dismay at being characterized as a profiteer in buying up Owens Valley lands. The Los Angeles papers had distorted his windfall of Rickey lands and cattle, Eaton said in an interview with the *Riverside Daily Press.* He had actually paid $21,000 of his own money for "a bunch of calves that have to be turned into beef," he said. About the only real money he had made out of the deal, he said, was "about $150" in egg sales from chickens left on the ranch. On the other hand, he said, the city had bought a ranch worth at least $1 million for only $425,000 due to his skills as a negotiator, and all he had received was "a little mountain land that would not be worth 5 cents without the calves."

Worse yet was the extent to which his reputation had suffered in a place he cared greatly about. Some hotheads in Bishop had been ready to hang him, Eaton claimed. "They even went so far as to buy the rope." But he was gratified by how things had turned out since. "I am now thankful that the people up there understand me," he said. "I have plenty of warm friends throughout the entire valley." So far as his treatment in the Los Angeles papers had gone, however, Eaton was "disgusted." As he told his interviewer, "I am going up to the ranch with my calves, and let Los Angeles severely alone."

In fact, Eaton made good on his statement, spending nearly two decades raising cattle and chickens in the valley, apparently devoted to enterprises that T. B. Rickey had been happy to rid himself of. Still, argument ensued as to the propriety of the ex–Los Angeles mayor's motives. To Mulholland, Eaton had acquired his Long Valley Ranch for no more than 20 percent of its actual value while acting as the city's agent, and, furthermore, he had known of the value of the property as a potential reservoir site from the beginning. Thus, he should be asking a far more reasonable price when the city was finally willing to negotiate. Eaton, on the other hand, believed that Mulholland opposed the sale simply on personal grounds and was making innocent people in the Owens Valley suffer because of it.

While the latter argument is sometimes still raised by Mulholland's critics, there are certain issues involved that would seem to go beyond personal feelings. Mulholland was always fearful of promising or delivering the city's water to places from which it might have to be reclaimed. Even if the city did not need all the water to be stored in a proposed Long Valley Reservoir in 1923, it could need it someday. And taking it back would be a vastly complicated, even impossible process, for once the water was employed for a specific use in the valley, prevailing law would make it difficult for Los Angeles to reclaim it.

A 100-foot dam costing little could be justified, but a 150-foot dam (especially one that in Mulholland's mind required the payment of an extortionate $1 million for acquisition of the lands) was simply beyond the pale. To put it simply—and as even critic William Kahrl agrees—Mulholland never believed that a Long Valley Reservoir could supply both the demands of the Owens Valley and the City of Los Angeles in the long run. It could possibly be useful as a source of hydroelectric power, but given the obstructionist stance of private power companies in the valley, that seemed unlikely as well.

Meantime, the steadily dropping water table had made pumping on city-owned lands farther south around Independence impractical, and the Public Service Commission redoubled its efforts to purchase more promising lands in the Bishop area. Finally, a group of property owners along the McNally Ditch, a sizable irrigation canal, agreed to sell their lands to the city for a price totaling more than $1 million. It might have marked the collapse of the Watterson Brothers' attempts to stand against the city, except for the fact that the headgates to the McNally Ditch were located well north of the aqueduct diversion point at Charley's Butte.

Even if the McNally waters were turned back into the river, the owners of all the irrigation canals in between the McNally and the aqueduct intake, including the sizable Big Pine Canal, vowed to simply suck that water through their own gates. The practice amounted to theft, but among local ranchers it became a running joke that it was certainly good of the city to spend $1 million to solve their water shortage.

For its part, the city devised an ingenious solution. Since the headgates of the Big Pine Canal lay on a spot where the Owens River took a slight jog to the east before turning again and continuing south, earthmoving equipment was sent out to a point just above those headgates. The intention was to dig a diversion canal across the narrow U of land and bypass the entry to the Big Pine Canal altogether.

When the city's intention became clear, a group of property owners hurried to the home of George Warren, Big Pine's representative for water dealings. "We need to get an injunction," one of the property owners shouted. Warren nodded. "You're right," he said. "We're going to get a shotgun injunction." With that, Warren led an armed group to confront workmen at the work site, and after a brief negotiating session the city's equipment somehow found its way to the bottom of the Owens River and the workmen were on their way back south.

It was a stirring victory, but one that proved short-lived. The city countered by reopening negotiations with property owners along the Big Pine ditch and by October 15, 1913, the largest part of Big Pine water was conveyed to Los Angeles, which had purchased 4,416 acres for a total of $1.1 million, or about $250 an acre, more or less the value of what prime irrigated land was selling for in the San Fernando Valley at the time. Though those who held out were never denied their historical share of irrigation water, the political situation in the Owens Valley had reached a tipping point. While some valley citizens consoled themselves with attempts to force city land agents into offering higher prices for lands, other residents were outraged by what they saw as stop-at-nothing tactics employed by an uncaring municipality intent on destroying life as it was known in the Owens Valley.

With the acquisition of additional water rights in the Owens Valley having eased the immediate water crisis somewhat, Mulholland turned his attention for the remainder of 1923 and early 1924 to campaigning for the Colorado water project, as well as to projects involving improvement of the existing system, including plans for a sizable storage impoundment to be created by the St. Francis Dam. As a vote on the Boulder Dam project approached, a series of articles appeared in the *San Francisco Call* in late March and early April, calling attention to what was described as "the tragedy of the Owens Valley." The writer, Court Kunze, married to a Watterson sister and former coeditor of the *Owens Valley Herald,* painted the citizens of the valley as being at the mercy of Los Angeles tyrants: Fred Eaton had been "ruthless" in his dealings with the valley people, Kunze declared, and Mulholland "has conceived himself in the role of a Pasha whose Armenian province was the Owens River Valley." Still, Kunze said, all could be made right if only a 150-foot dam was built at Long Valley.

It can only be imagined what Mulholland thought were the true motives of the writer, but the pieces proved popular and were reprinted widely throughout the state. Whether the articles were responsible for what happened next is impossible to determine, but one thing is certain: on May 21, 1924, the first salvo in what has been immortalized as the "Owens Valley Water Wars" was fired.

LET THE BOMBINGS BEGIN

AT ABOUT 1:30 IN THE MORNING, THE QUIET DESertscape in the Alabama Hills north of Lone Pine was rocked by an explosion. A 500-pound charge of dynamite blew away a 100-foot section of the concrete aqueduct, loosing river waters across the adjacent scrub lands, and bringing Mulholland quickly to the scene to survey the damage. The blast drew renewed attention to charges levied by the *San Francisco Call,* and Mulholland was reviled in Owens Valley papers with renewed vigor as "King of the Home Destroyers." Though a reward of $10,000 for the saboteurs was offered, and the perpetrators were widely assumed to be disgruntled valley residents, no one was ever caught.

Reports of the incident reminded readers that there had been negotiations between property owners in the Owens Valley and the city since 1913, and that there were about 5,000 residents in the more northerly portions there who claimed to be dependent for their livelihoods on Owens River water. An unidentified source within the city's power bureau acknowledged the hardships that the loss of water might effect upon the valley's agricultural economy, but chalked it up as a question that had to be answered by the dictum of the greatest good for the greatest number. "Individual owners can be compensated," the source said, "but there is an essential conflict between the life of the aqueduct, which is the life of this city, and the agricultural prosperity of the Owens River Valley."

Mulholland told reporters that the real force at work in the Owens Valley was not desperation but greed. "There is a movement afoot to force us into acquiring rights in the Owens River Valley at prices which we have not been willing to consider," the Chief said. "The real grievance in the valley is caused not by what we have bought but what we have declined to buy, and the real terror in the valley is not that we will acquire more but that we will turn to the Colorado River [to meet future needs]."

In August, the Public Service Commission released a report showing that the city had become the owner of fully half of the irrigable land in the Owens Valley—about 27,000 acres—and that the total valuation of property in Inyo County had risen from about $2 million in 1905 to more than $10 million. Rather than "ruining" the valley, the report said, the city was actually improving it. In 1923, the city would be paying almost a quarter of all taxes received in the county, an amount equal to the total tax revenues in 1905, or $70,000.

None of this did much to allay passions in the Owens Valley. On Wednesday, August 27, Lester Hall, a local attorney suspected of helping wildcat property owners sell to the city, dropped into a Bishop restaurant for dinner. Scarcely was he seated at the counter than a group approached. Hall, who'd already been threatened a number of times, reached for the pistol he carried, but he was too late.

One man snatched his pistol away and another put an arm around his throat. In seconds, Hall was dragged through the restaurant doors and flung into the back of a waiting car that sped off into the desert south of town.

Though no one spoke at first, there was no mistaking the meaning of the coil of rope that lay on the floorboards of the car. When the car, now part of a convoy, finally pulled off beneath a cottonwood tree on a deserted stretch just above Big Pine, about fifteen miles from Bishop, Hall was ready for the worst.

He was dragged out of the car, and the rope was tossed over an overhanging limb. It was all made worse by Hall's having recognized most of the men in the group that surrounded him, more than twenty strong by now. "For God's sakes," he said. "I'm 52 years old. I haven't done anything to be ashamed of. If you do this, it'll be on your consciences the rest of your lives."

As a last resort, Hall, a member of a fraternal organization that included most of the men who surrounded him, made a gesture of distress meant to rally support from fellow members. There was a murmur from the group, and its leader stepped aside to confer with a pair of his lieutenants.

After a moment that seemed for Hall to stretch forever, the ringleader, a man Hall had known for most of his life, approached. "We've told you to leave the valley," the man said. "You haven't, but you'll have another chance. Are we clear?"

They were more than clear. Hall was tossed back into the car and taken to Big Pine to the home of local Chamber of Commerce president George Warren. The next morning, C. C. Collins, the sheriff of Inyo County, came by to escort Hall to the railway station from which he made his way to Glendale, where he would spend the remainder of his days.

When the story hit the papers, Los Angeles reporters approached Mulholland, who was scheduled to make a fence-mending trip, along with members of the Public Service Commission, to the valley the following Tuesday. Was the Chief thinking of cancelling his plans, reporters asked, given what had happened to Hall? Mulholland snorted. He'd been receiving threatening letters from the Owens Valley for years, he said, and he'd been assured recently that he would be killed if he went up there next week. If he let such things influence his behavior, he would never get anything done. "They wouldn't have the nerve," he added. He was going to the Owens Valley and that was that.

At this Owens Valley summit, which took place on the twenty-year anniversary of the now sixty-eight-year-old Mulholland's first trip to the region with Fred Eaton, George Warren, Hall's savior and head of the Associated Chambers of Commerce of Inyo, told the visitors, "We know if you take away the water from the land you have already bought and give Big Pine nothing in return, we are ruined," he said. "We expect a square deal from the City of Los Angeles and believe we are going to get it." Warren was referring to a hoped-for reparations settlement to Owens Valley businessmen who had been hurt by the exodus of settlers from the valley as a result of the city's land buyouts. While property owners might indeed be satisfied to take the city's money and run, Warren pointed out that local businessmen had no such option.

In fact, by the city's own figures, the Bishop area suffered a 20 percent decrease in population in the 1920s, causing six elementary schools to close and another six to be consolidated. Though the city began a program of leasing some farms back to tenants, the terms were for only five years, discouraging anyone with long-term interests and limiting most leaseholds to the planting of hay and alfalfa. Only forty-three such leases had been granted, and some businessmen and professionals complained that they had lost as much as

50 percent of their income. The hope was that the city would reimburse business owners for the present value of their property should future damage result from turning off water to ranches in the area.

In addition, a consortium of local property owners urged that the city simply complete the purchase of all private land in the Owens Valley and put an end to all the uncertainty. "Our property values have depreciated to the vanishing point, insofar as any purchaser except the City of Los Angeles is concerned, and outside capital is no longer available," a statement prepared by local banker Wilfred Watterson read. "Renewal of mortgages is no longer possible; the Federal Land Bank refuses absolutely to grant further loans; fire insurance companies have withdrawn from the field; our rural population is decreasing; farms are deserted; former homes untenanted are left to the ravages of the elements; school districts have lapsed; others are going; the number of teachers of course has decreased; and some of our best citizens have forsaken the valley on account of its problematical future."

Mulholland assured the group that the city was looking elsewhere for its future needs, including the Colorado River and a second aqueduct to Mono Lake, which drained the Sierra in the next valley north of the Owens, and he stated his sympathies for local residents. "The people are entitled to justice," he said, "and justice they shall have." In the end, it was determined that the Big Pine committee would put a specific reparations proposal together and submit it to the City of Los Angeles at a date in the near future.

For a short period there was hope that a compromise might be worked out—the valley would have been willing to settle for a buyout of $8 million for all remaining private lands and a settlement of about $150,000 for businessmen, but Mulholland was opposed to paying that high a price for the land, owing to the fact that a goodly portion of it held no water rights and was thus virtually worthless. In his eyes, he was not obstructing the "justice" he had spoken of in Bishop, but was defending Los Angeles against unscrupulous individuals who wanted to profit at the expense of city taxpayers. The city's counterproposal was to ensure a continuing water supply sufficient to irrigate 30,000 acres of valley land, perhaps 10 percent more than was currently supplied—but it was a proposition that residents there found "ridiculous."

The matter came to a head on November 16, a little more than a month before the dedication of the scenic Mulholland Highway linking Calabasas with the Cahuenga Pass near Hollywood. On that Sunday morning, a caravan of automobiles carrying about sixty Owens Valley citizens left Bishop to travel nearly sixty miles to the Alabama Gates just above Lone Pine. The men advised the guards at the intake that their presence was no longer required, a suggestion that was readily adopted. In short order, the group had swelled to more than a hundred, and cheers erupted as the gates were opened and the waters turned out onto the valley floor below. For the first time in a decade, the Owens River was once again coursing toward the flats where a vast lake had once stretched.

When Sheriff Collins arrived at the scene, his request that the men leave had no effect, and Collins saw little hope of forcing the issue. Likely motivated by political expediency as much as by common sense, Collins telegraphed a plea to Sacramento, asserting "the party will disperse and bloodshed be averted only by arrival of State troops." For his part, Governor Friend William Richardson said he would sleep on the request, suggesting that Collins could try a bit harder to handle the situation. While these political veterans jostled, the entire flow of the aqueduct, at least a full 5 million gallons an hour at that time of year, was being turned out on the desert floor at a cost to the city of about $10,000 a day. Though there was a three-month supply of water stored behind the Haiwee Dam, Mulholland quickly assured the citizens of Los Angeles, the situation was obviously serious.

Assistant engineer Harvey Van Norman, who had been involved in trying to broker a compromise with valley interests, told reporters, "The group which seized the aqueduct today does not represent the majority of the people of the Owens River Valley," but anyone who chanced by the scene might have thought differently. The occupying group quickly grew to 350, including a number of women, who came to cook, and their children, who soon were assembled into choirs and makeshift drumming bands. Barbecues were organized and, according to one report, even Sheriff Collins was spotted with a heaping plateful.

Another report claimed that the town of Bishop was deserted, with a hand-painted sign planted at the city limits stating, "If I am not on the job you can find me at the Aqueduct." Sheriff Collins claimed that he was not taking the situation lightly and issued a second call to the governor for troops. There were easily a hundred sawed-off shotguns stashed among the occupiers, Collins asserted, and unless outside help was sent, he said, "I believe dynamite will be used on the Aqueduct and the water supply of Los Angeles will be in the greatest peril."

By Wednesday, the leaders of the occupying force were themselves calling for the governor to send troops, telling reporters that they would "disperse without incident" if state militia came in, and the Inyo County district attorney embarked for a conference in Sacramento with Governor Richardson. In response, the governor wondered what kind of dire situation it was when the rabble-rousers themselves were the ones calling for the quashing of their efforts. Finally, he agreed to send State Engineer W. F. McClure to investigate the situation, but McClure's most notable action was to join in at a barbecue that fed a crowd of 700.

The standoff was finally resolved on November 21, five days after it began, when Los Angeles banker J. A. Graves proposed the formation of a mediation panel consisting of three state judges to resolve the claims of Owens Valley property owners and business-owners. With that, the leaders of the takeover agreed to disperse. The spillway at the Alabama Gates was closed, restoring the flow of the aqueduct water, and the occupying citizens went home. In short order, the Wattersons delivered their final proposal to the mediation committee: that the city pay $12 million for the remaining irrigable lands and $5.3 million in reparations claims. J. A. Graves took the Watterson demand under consideration, though he did tell reporters that one of the main questions in the committee's deliberations would be whether any monies the city agreed to pay would go to farmers of the valley or to a "junta of capitalists" who did their farming through the newspapers and simply wished to cash in on speculative investments by dint of virtual extortion from the city.

Debate on both the issues of additional land purchases and reparations dragged on for years, with the Public Service Commission rescinding its proposal to guarantee the maintenance of 30,000 irrigated acres and resuming a policy of piecemeal acquisitions, though the city did agree to the appointment of a three-person panel of valley officials to assess the value of properties. Using the figures of this board, the land purchase program continued until May 1, 1927, by which time the City of Los Angeles was the owner of about 225,000 acres, or 80 percent, of the privately owned farmland in the Owens Valley.

The 1925 session of the California legislature approved a measure calling for reparations to Owens Valley residents where city liability could be proven. As a result, depreciation claims came in from the owners of town lots, homes, and buildings totaling nearly $2 million. The owners of sixty-seven businesses claimed losses, including surplus equipment, totaling almost $700,000. The claims of various doctors, dentists, mechanics, bank clerks, stenographers, bookkeepers, electricians, blacksmiths, barbers, and beauticians totaled something over $125,000, and thirty-five "Indians" said that they had lost $25,000 in wages they otherwise would have earned as farm laborers. In total, the claims were more than $2.8 million, though the Public Service Commission declined to make any immediate payments.

With the disbanding of the Graves Commission and the intransigence of the Public Service Commission, tensions escalated in the valley and in 1926 a series of bombings once again shook the line, one of which took out a section of the aqueduct near Lone Pine, not far from the Alabama Gates, on May 12. Mulholland did not seem greatly concerned by the actions, characterizing the Alabama Hills blast as "just another gesture" in an attempt to intimidate the city. "They do it now and then up that way," he said. "The dynamite is set off one night and we get a boxful of reparations claims the next morning."

At about the same time, and while a renewed fight to enforce the payment of reparations was being debated in the state legislature, R. P. Del Valle, president of the Board of Water and Power Commissioners, issued a statement that called the portrayal of the Owens Valley as the "Land of Broken Hearts" into question. Charges that the valley had suffered as a result of the city's land purchases, many made at as much as four times the assessed market value, had never been substantiated by concrete evidence. On the contrary, Del Valle said, there had actually been "an increase in bank deposits, growth in railroad freightage and building construction," over the period. Much of the controversy, Del Valle contended, was the work of private electrical power interests who were still battling the city in an effort to maintain their monopoly and who had conspired with others in the valley eager to extort payouts from the municipality.

When the State Senate tabled the new reparations measure on April 27, the response from the valley was forceful, though not in terms of debate. Another explosion on May 27 took out 450 feet of the No Name Siphon in the Grapevine Division, an action that would disrupt water flow to the city for about three weeks, while another damaged a control gate near Big Pine on the following day. City officials decried the actions as nothing less than a blackmail attempt by valley interests, and when reporters asked Mulholland if he had any comment to make, he assured them that very little of what he had to say could be printed.

Release of a battery of statistics supporting Del Valle's contention that the economy of the Owens Valley had improved significantly as a result of the city's investments there produced further bombings by way of response, one on June 5 near the junction of the aqueduct with Cottonwood Creek, close to Lone Pine, and two less significant blasts on July 15 and 16. After conferring with various elected officials from Inyo and Los Angeles, including William B. Mathews, Governor Clement C. Young issued a statement condemning the bombings and calling upon law enforcement to apprehend those responsible. To put an end to the strife, Young proposed that ranchers and other valley citizens should immediately bring a series of test-case suits against the city under the provisions of the Reparations Act of 1925. The city would then settle all outstanding reparations claims based on the outcomes.

Young's proposal might have put an end to the matter, but before any action could be taken, affairs in the Owens Valley took a completely unexpected turn. Rumors had been circulating for some time about the health of the Watterson Brothers' five valley banks, and on August 2, a state bank examiner appeared to inspect the books. As it turned out, there was a vast discrepancy—about $500,000—between the cash on hand and what was owed depositors. All of it had appeared to have gone to support various private enterprises owned by the Wattersons.

By August 4, the banks were shuttered and by the end of the month Wilfred and Mark Watterson were arraigned. The two would later claim that their only motivation had been to keep the valley's economy alive, but they would be convicted on charges of embezzlement and sentenced to San Quentin. It was a stunning blow to the valley community, many of whose citizens lost their life savings. With the disgrace and fall of the Wattersons, organized resistance to the City of Los Angeles was at an end.

FAILURE

THE END OF THE OWENS VALLEY WATER WARS PROVED A great relief to William Mulholland. Though he lamented the fallout upon the innocent from the Wattersons' perfidy, he would go to his grave certain of one thing: his own dealings with property owners in the Owens Valley had always been on the square.

Shortly after the Watterson Brothers closed the doors of their five banks in the valley (two in Bishop, one in Big Pine, one in Independence, one in Lone Pine), Wilfred Watterson issued a statement that depositors would be able to recover their funds "when the Owens Valley reparations claims against the city of Los Angeles have been paid." When word of this reached Mulholland, he was nearly apoplectic.

"Astounding," Mulholland described the claim. "What has the reparations business to do with the financial condition of Watterson's banks?"

"The first statement made by these valley bankers when they closed their doors was that their banks had been affected by 'frozen' loans on ranches," Mulholland said. "The farming interests have never even made any reparation claims against the city."

He pointed out that the city had made about $12 million in land purchases in the Owens Valley and in many instances the properties were mortgaged by the Watterson banks. In every instance, Mulholland said, the sellers had received cash in excess of the amount of the mortgages and the bank could have—certainly *should* have—gotten its money out at that time. (The astonishing fact would come out at the bankers' trial that many mortgages—including one on a well-known piece of property—were simply never cancelled by the Wattersons, even though they had collected the payouts. Thus, the mortgagees were left still responsible for the notes even though the bank had been paid.)

Mulholland disputed the notion that the city had devastated the valley economy and suggested that the reparations group had engaged in a long propaganda campaign to bolster hopes of a large payout from the city. As to the Wattersons' claims that the city was somehow responsible for the banks' condition, Mulholland said he was happy to let state bank examiners make that determination. Meantime, he said, it was his sincere—though ultimately forlorn—hope "that the people of the valley who have money in those banks will be able to recover it."

There had been other matters for Mulholland to deal with throughout the stormy period, not all of them distressing. The new Mulholland Highway was dedicated at its Calabasas terminus on December 27, 1924, when the Chief smashed a bottle of aqueduct water over the gates and inserted a golden key to open the road to traffic. A rodeo featuring Tom Mix and a cadre of Hollywood cowboys ensued, followed by a caravaning of dignitaries along the twenty-five-mile twisting road that crossed the now-legendary canyons in the hills—Sepulveda, Beverly Glen, Benedict, Coldwater, Laurel—all the way to the Cahuenga Pass near the Hollywood Bowl. The event was marked with a huge parade in Hollywood, and Mulholland later spoke to about 10,000 people gathered at the bowl, the landmark venue that opened in 1922 but that would not permanently install its iconic shell until 1929.

Mulholland admitted that he had been advocating such a scenic road from the days of the aqueduct's arrival in the San Fernando Valley, but it had only recently become a reality when a group of private investors owning 10,000 acres along the route underwrote the $1 million bond issue for its construction. "Persons having learned the advantages of living above the fog and turmoil, smoke and congestion of the city, would flock to the hills," one of the developers predicted, admitting that the 70,000-acre area had long been considered "worthless" owing to the difficulty of the terrain.

On March 17, 1925, Mulholland was again honored when the 200-foot high, 1,000-foot wide dam in Weid Canyon holding back the scenic Hollywood Reservoir was dedicated in his name. At the ceremony dedicating the structure, Mulholland remembered working nearby to quarry rock for the city's jail forty-three years before. His colleagues had been ribbing him that some of the fifteen impoundment dams he had constructed as part of the water system resembled old women's aprons. "But in this job I think I may take a little pardonable pride." In fact, the gracefully curved dam, the first concrete dam Mulholland built, still stands as a striking example of its kind. Surrounded by a broad footpath, it remains a popular destination for joggers and bikers. News stories carried what had become a characteristic footnote to the completion of a Mulholland project—typical costs for comparable structures in other cities ran at ten to twelve dollars a cubic yard. Mulholland Dam cost but six dollars per yard.

Mulholland also spent considerable time in the 1920s working vigorously on behalf of the Colorado River Project, which he saw as the only long-term solution to the city's future needs. It was a foregone conclusion that the city's population would pass the 1 million mark in the 1930 census (it would be nearly 1.25 million, in fact) and that of the county was already nearing 2 million. On February 29, 1928, he appeared before a group of Southern California municipal representatives gathered in Long Beach to discuss the implications of the state's approval of the Southern California Metropolitan Water District, formed to distribute Colorado River water to San Diego, Los Angeles, and Southland communities in between.

Mulholland explained to the group that the key to the entire project's viability was the construction of a "high" dam at Boulder in order that sufficient electricity would be produced to pump the water up over the intervening mountains. The Colorado River Aqueduct would be 260 miles long and carry enough water for 7 million, he told listeners. Los Angeles would receive 1,500 cubic feet per second of that flow, or about three times what the Owens River was providing, but it all hinged upon the building of a properly sized dam.

It is ironic that much of Mulholland's time was being devoted to questions of dam construction, particularly given what was about to happen. Not only was the Boulder Dam on his mind, but he had also been called to the site of the new impoundment dam in the San Francisquito Canyon on March 12 by dam keeper Tony Harnischfeger. The St. Francis Dam was the most recent of nine impoundment structures Mulholland had built or enlarged for the system in the 1920s, including the Lower Franklin, Stone Canyon, Encino,

Sawtelle, Ascot, and Hollywood/Mulholland Dams. He hoped to be able to store a full year's water supply in the new reservoirs, Mulholland wrote in his "Twenty-fourth Annual Report." Largest and newest of them all was the St. Francis Dam, which could store 32,000 acre-feet of water, about half the total of the entire group.

As fate would have it, the site of the St. Francis Dam was not Mulholland's first choice. He had, in fact, planned "the big one" for Big Tujunga Canyon, but when the department began condemnation proceedings on lands there, property owners initiated a series of legal challenges meant to drive up prices. In disgust, Mulholland ordered proceedings in Big Tujunga stopped in favor of a second choice, one he had long identified as a potential dam site. During his surveys for the original aqueduct route through the San Francisquito Canyon, Mulholland had identified the area lying between Power Plant #2, well down in the deepest folds of the canyon, and the site of Power Plant #1, at the northerly head of the broad valley upstream, as a natural place for a reservoir. It would be a relatively simple matter to place a dam at the bottom of that valley, at the spot where the canyon began to narrow, and there was plenty of room to store waters behind it. More importantly, there were few property owners to contend with.

The structure was designed as a curved concrete gravity arch structure, much like the Mulholland Dam in the Hollywood Hills. Though it was never documented, the choice in design was probably made because of the lack of sufficient clayey material in the surrounding soils, thus ruling out the use of hydraulic sluicing that Mulholland had used in building most of the other structures. A dam such as the St. Francis is called a "gravity" dam because its basic strength in holding back the waters pressing against it comes from its dead weight, enhanced by the width of its "sole" pressing against the ground at its base.

One aspect of its design that would prove controversial was the fact that the height of the dam above the creek bed it crossed was planned to be 175 feet. During the actual pouring of concrete, however, it was decided to raise the height to 195 feet, so that as many as 38,000 acre-feet of water could be stored. As experts have since noted, the raising of the height of a gravity dam without a corresponding increase in the thickness of its base would be dangerous.

There was also another issue involved, one that was only then beginning to be appreciated in the civil engineering community—the impact of uplift forces at work on a monolithic structure such as the St. Francis Dam. Once water is impounded behind a dam, the great weight of that water presses not only forward, but downward, exerting great pressure on underground water unavoidably percolating beneath the bottom of the dam itself. Today, no such dam would be built without great attention to the drilling of "uplift relief wells" beneath the base, but some analysts contend that such science was not fully understood at the time.

Another factor to be considered when estimating the stability of a concrete dam is the relative porosity of its concrete. To the degree that concrete can become saturated by water, it actually becomes buoyant and the effect of its weight is counteracted accordingly. Some experts have theorized that owing to the nature of the aggregate taken from the nearby soils to mix with the cement, the porosity of the structure was substandard and its weight was actually about 7 percent lighter than that of comparable structures.

In any event, construction on the St. Francis Dam was completed in May 1926 and by May 1927 impounded waters were at 177 feet, three feet below the spillway. During the time of filling, cracks appeared in the downstream face of the structure, which Mulholland identified as "transverse contraction cracks" to be expected as part of the curing process. The cracks were filled and sealed to prevent further seepage, and with the lowering of the water level during the summer months, all seemed well.

Then, in February 1928 as the spring runoff again raised the water levels, some leaks began to appear on both the east and west sides of the dam's foundation. In addition, the cracks that formed the previous year reopened, and a sizable leak (four to five gallons per second) opened in a concrete wing extending out westward from the crest of the dam. To avoid erosion, Mulholland ordered crews to install an eight-inch underdrain below the wing that carried the leaking water back toward the canyon where it splashed down the face of the dam, giving the impression that it was gushing from the main structure.

On Monday morning, March 12, the waters were overrunning the dam's spillway, and operators at Power Plant #2 downstream opened gates that dumped aqueduct waters from the system into the normally dry bed of San Francisquito Creek. Of course, dam keeper Tony Harnischfeger had by that point summoned Mulholland and Van Norman to inspect what he thought was an impending blowout on the west flank of the dam. Mulholland and his assistant arrived about 10:30 and spent the next two hours inspecting the various leaks and seeps. As to the new leak, Mulholland pointed out to Harnischfeger that the water coming from under the dam on the west embankment was clear. It did not pick up its muddy color until it ran across the steep access road that had been recently cut into the hillside there. That meant that the water was either leaching out of the concrete or was being forced up from the underlying aquifer. Neither situation was cause for alarm, Mulholland said. The dam was not in danger of being undermined.

Though he had reassured Harnischfeger, troubling thoughts could not have been far from Mulholland's mind in the aftermath of that visit, as his daughter Rose was to attest. When the telephone rang shortly after midnight in the Mulholland home, it was Rose who picked up the call. Harvey Van Norman was on the line, offering a terse explanation of what had happened. As Catherine Mulholland retells the story, "She went to her father's bed and awakened him with the news. As he rose and stumbled toward the phone, she heard him repeat over and over, 'Please God. Don't let people be killed.'"

Mulholland and Van Norman were rushed to the scene in a car driven by Mulholland's son Tom, who was still living at the family home near Third and Western. Inside two hours following their receipt of the news, the men, using back roads through Bouquet Canyon, were at the scene of the disaster, where only one narrow central section of the broken dam was still standing, looming against the night sky like the exclamation point of doom.

It was quickly obvious to Mulholland that his prayer had been a hopeless one. Nearly everyone who had been in the canyon or anywhere close to the stream beds of San Francisquito Creek and its junction with the Santa Clara River had died, and the bodies that could be found were already being carried to temporary morgues at Saugus, Newhall, Fillmore, Piru, and other communities that lay near the route of the water's fifty-five-mile rush down the river valley to the sea. According to a coroner on the scene, the majority of the victims he'd seen had not been drowned but crushed by cascading boulders or flood-driven timbers and pieces of machinery.

One employee at Power Plant #2, Lyman Curtis, was awakened by the rumble of the collapse and sensed what was coming. He shook his wife awake, thrust their three-year-old son Danny into her arms, and sent her scrambling into the nearby hills. Satisfied that she would be safe, Curtis turned and ran inside their cabin for the couple's two daughters. Mrs. Curtis made her way to the top of a rise and paused to wait for her husband. She turned to see him hurrying out the door of the cabin below, the two children in tow. And then the waters covered everything. She never saw her husband or her other children again.

Another woman, Ann Holzcloth, spoke to a reporter from a cot in an emergency hospital set up in the Santa Paula schoolhouse, more than forty miles downstream from the broken dam. She had been asleep in her home in the Santa Clara River Valley when, without warning, floodwaters slammed into the house and swept it from its foundations. "The baby was sleeping with me," she said. "I clutched him tight as we were swept out on the water in the dark."

She managed to grab onto a floating timber from the wreckage of her house, Ms. Holzcoth said. "With my other arm, I held the baby out of the water the best I could. I know that he was alive when we hit a whirlpool."

In the next moment, she said, the swirling water wrenched the child from her grip and threw her in an opposite direction. "I landed on dry land," she sobbed. "Why did I have to live?"

Reporters described other horrors from the scene, including the "pitiful figure of a woman, huddled in a vivid red sweater, wringing her hands." She identified herself as Mrs. Russell Hallen, and explained that her young daughter had been living in the San Francisquito Canyon with her grandmother. "Right over there," she told a reporter, pointing to a place where a large cottonwood jutted from an otherwise flattened and featureless plane of silt. Not only were the tree's leaves gone, but the bark had been stripped from its trunk and limbs, leaving it glistening like a fan of bones.

A man wearing a long fur coat was spotted staggering through the mudflats by the still-running creek, tearing through clumps of debris calling out a pair of names in endless succession. He was Jimmy Errachow, the man said when a reporter stopped him for a moment, but he did not have time to talk. His wife and his baby were still missing and he had to find them.

Thirteen-year-old Luis Rivera, who lived on a ranch downstream near Castaic, was awakened by the roar of the coming waters and woke his father. "The water is coming," young Luis cried.

"It is only the rain," his father told him and went back to sleep. Luis was not pacified. He woke his younger sister and pulled her out of bed and into the hills despite her protests. They left behind their father, their mother, and an older brother, all of whom died in the flood.

One hero on the scene was identified by workmen among those at the Southern California Edison tent camp set up at Piru, about twenty miles down the Santa Clara River from the dam site. A man named Locke, the camp watchman, had seen what was hurtling down the river bed toward them and raised the alarm. He ran from tent to tent, calling on the sleeping workmen to flee. "I got out because of Locke," one of them said. "He was still running from tent to tent when the water took him away." In all, 84 of the 177 men in the camp were killed, but it was agreed that had it not been for Locke's efforts, the toll would have been even worse.

A letter to Catherine Mulholland from Daisy Orton, a family friend living with her husband in Fillmore, about forty miles downstream from the dam, lends some sense of the fear and confusion among the residents the night of the collapse: "The fire bell at 2 o'clock woke us & we could keep hearing an awful roar," Ms. Orton said. Though the Ortons were warned that they would have to flee, the water rushing through Fillmore crested in the block just below their home. "We didn't go back to bed," she said, "but built a fire in the fireplace and stayed up. It was sure a terrible night and I don't believe I will ever forget that awful roar of the water. Who would ever have thot that such a calamity could have overtaken us here."

In the aftermath, Ms. Orton described the difficulties in moving relief and rescue workers onto the scene owing to the collapse of bridges and highways. "The river is absolutely clean of trees," she said, and all the houses close to its bed had been destroyed or taken off their foundations. Her husband, Luce, was gone for the next two days and nights with search parties looking for victims. They had found fifty at the time of her writing on March 14, "not drowned—but battered and bruised—but don't show anguish—so that probably they were taken in their sleep and didn't know what had happened." Ms. Orton added that there appeared to be plenty of outside aid coming in, "and as far as we can hear—everything is being done that can possibly be done."

Given that census statistics were not then what they are today, a precise death toll has never been firmly established, though it is generally agreed that somewhere between 400 and 600 people died in the disaster. Twenty families were completely wiped out. The only dam collapse that took more American lives was the epic failure of the South Fork Dam and resultant Johnstown Flood of 1889, which killed more than 2,000. Combined with the destruction of roads, bridges, buildings, and as many as 1,000 homes, the loss in both lives and property (estimates ranged from $15 million to $50 million), the St. Francis Dam collapse is considered to be among the worst civil engineering disasters in US history and ranks behind only the San Francisco earthquake and fire among the catastrophes that have struck the state.

As to relief efforts, most work centered on restoring roads, rail lines, and utilities. There were few injured, as it turned out, for victims had either been killed or escaped without harm. As one pilot who flew over the scene remarked, "No rescue work is needed, but you will have to hunt for bodies."

The efforts of local civic and veterans organizations were reported as exemplary as were those of national organizations such as the Red Cross. Three hundred city policemen were dispatched to the region to help maintain order, the Safeway grocery chain began daily shipments of 2,000 loaves of bread to field kitchens, and President Coolidge pledged to send troops and aid in Red Cross efforts.

On March 19, the City Council, prompted by recommendations from Water and Power commissioners, approved an immediate appropriation of $1 million to compensate victims of the tragedy and established a joint committee to review claims. According to the committee's report, about $915,000 in death and injury claims were paid out by July of the following year. By March 24, the city had also arranged with the General Contractors Association of Los Angeles to oversee the cleanup and rebuilding of homes in the area, and within two days, 1,600 workmen and 95 pieces of heavy equipment were at work there.

With the immediate shock beginning to lift, attention turned to questions as to the cause of the disaster. Elwood Mead, chief of the US Bureau of Reclamation, was appointed by the City Council to head a board of inquiry, and Governor Young ordered the state's director of public works to name a board of investigating engineers. Los Angeles District Attorney Asa Keyes also announced a criminal investigation into the matter that would begin with an inquest by the coroner.

Speculation was immediate that the composition of the rock on the opposing canyon walls abutting the dam was to blame. "Undoubtedly that is what caused the St. Francis Dam to break," said Ventura County chief engineer Charles Petit. "The decomposed rock at the sides, weakened by the water, gave way, and then the structure went out." Other Ventura County officials, though admitting they were not geologists or engineers, chimed in, claiming that even a layman could see that this rock was unsound and could be reduced to pulp when mixed with water.

For his part, Mulholland speculated that a landslide just above the dam might have caused the disaster. "Something terrific happened to break that huge mass of concrete into bits," he said. In the end, his words would prove prescient though not exactly in the way the Chief had intended.

As relief and rebuilding efforts continued in the Santa Clara Valley, where about 1,500 remained housed in tent cities, Mulholland appeared before the Board of Water and Power Commissioners to make a special request. As his presence might cause an embarrassment to the department until the investigations were completed, perhaps it would be best for him to take a leave of absence, the forty-year veteran of the water works said. The request, said reporters, brought tears from members of the board, most of whom had known Mulholland for many years. In the end, Mulholland's request was denied. "The board hereby declines to grant said request and urges the Chief to remain on the job he has so faithfully filled for half a century."

Though Mulholland would have been heartened by such shows of support, he was also well aware of the swell of countervailing views—such as that expressed by a sign erected in the front yard of one flood-damaged Ventura County home: "Kill Mulholland."

It was the first disaster involving any project overseen by Mulholland in a long and storied career. As he would say during his testimony, "I've built nineteen dams and have been consulted on nineteen more and this is the only one where anything went wrong." Still, the toll on the man was evident at once to those around him. "Chief Engineer Mulholland was a pitiable figure" as he appeared before the Water and Power Commission on the afternoon following his inspection of the disaster site, a *Los Angeles Times* reporter noted. "His figure was bowed, his faced lined with worry and suffering, [and] his voice was broken," the piece continued, concluding with the observation, "The tragedy of the people in the canyon and the Santa Clara Valley is the tragedy of William Mulholland."

In a 1988 interview with Catherine Mulholland, a secretary in the DWP legal department, Lillian Darrow, recalled Mulholland as having been before the disaster "such a sweet man." But immediately afterward, Darrow says, "He went down. His face aged twenty years."

On March 21, the coroner's inquest began, with more than forty witnesses called for the purpose of determining any criminal behavior in the matter of the dam's collapse. At the same time, the investigations of Elwood Mead and the Young Commission got underway, with the charge of determining the engineering flaws or causes attendant to the failure.

At the inquest, the seventy-two-year-old Mulholland's approach to the stand was described as "feeble," and questioning had hardly begun when he broke in with the remark, "On an occasion like this, I envy the dead."

When examination by the district attorney resumed, Mulholland carefully recounted his visit to the dam site on the morning prior to the disaster and reasserted his belief that nothing he saw that day was out of the ordinary. All dams leak, he insisted, but the St. Francis Dam leaked less than most. Nor, he said, counteracting rumors to the contrary, was there long-standing suspicion of dangers with the structure among any employees of the water department. "Had I such a suspicion or the slightest idea of the kind," Mulholland said, "I would have sent a Paul Revere down that valley. It never occurred to me that it was in danger." As engineer Clark Keely would recall many years later, he and a number of his coworkers from Power Plant #1 had in fact picnicked atop the dam the very day of Mulholland's visit.

Mulholland also declared that the base and sides of the dam had been anchored sufficiently, with topsoil removed and trenches cut into bedrock to a depth of thirty feet in some places. Mulholland also dropped hints that he believed that sabotage of the sort that had plagued the aqueduct project in the Owens Valley over the past years might have been involved, though he would not say so directly. When the district attorney asked if he had any explanation for the dam's failure, Mulholland said, "I have a suspicion, but it is a very serious thing to make a charge that I don't even want to utter without having more to show for it."

Following a trip by the inquest's jury to the dam site and the hearing of several other witnesses, Mulholland was recalled to the stand on March 27. As he was explaining that a number of holes had been bored in the abutting formations to be sure that water would not wash away the anchor points of the dam, Mulholland was interrupted by one member of the jury who wanted to know why he was so certain that his choice of a site for the St. Francis Dam was justifiable. Mulholland fixed his questioner with a stare. "I am willing to take my medicine like a man," he said. "I have nothing to conceal and will be the first to help you in any way possible to determine the cause of this disaster." Then, at the conclusion of his testimony, he uttered the words that essentially defused the entire inquiry. "Fasten it on me if there was any error of judgment, human judgment. I was the human."

Finally, on April 12, following consideration of reports by Mead and his consulting engineers, the jury rendered its verdict. The choice of the site for the St. Francis Dam was ill advised, the panel concluded, primarily because the nature of the foundation material at the west abutment of the dam was faulty. That material was subject to saturation by the impounded water and over time had begun to decompose at the place where the west wall of the dam was anchored. It was the west wall of the dam that had first given way, the jurors concluded, leading to the collapse of the rest, although Mead had identified issues with the east wall as well. Though that wall had appeared to be sound, it was in fact a mica schist formation, as prone to fracture as the gneiss rock Mulholland was once so happy to encounter in the tunnel below Lake Elizabeth.

The coroner reported that there were in fact two errors in judgment that had resulted in the dam's failure: The first was essentially Mulholland's, owing to "the very poor quality of the underlying rock structure upon which (the dam) was built and to the fact that the design of the dam was not suited to inferior foundation conditions." The second was more the fault of the Board of Water and Power Commissioners, for their lack of good judgment. It might have been understandable, given Mulholland's exemplary record of achievement, but Mulholland in truth had little practical experience in building concrete gravity dams, other than the Hollywood Dam in Weid Canyon. The board had thereby erred in allowing so great a responsibility to rest solely on one man.

More significantly, the jury had found "no evidence of criminal act or intent on the part of the board of waterworks and supply or any engineer or employee in the construction or operation of the dam." There would, as a result of these findings, "be no criminal prosecution of any of the above by the District Attorney."

FORGET IT, JAKE. IT'S CHINATOWN

FAMILY MEMBERS SAY THAT MULHOLLAND MAY HAVE felt only a grudging appreciation for the coroner's determination not to recommend that criminal charges be filed. In a 1978 interview with Catherine Mulholland, the Chief's nephew and namesake William Bodine Mulholland said that while his uncle appreciated the explanations offered by Elwood Mead and others for the failure of the St. Francis Dam, he was never convinced by them. "He had as much knowledge of those kinds of things as anybody in the world, and he used every bit," the younger William said, insisting that his uncle remained ever uncertain of the reasons for the dam's failure. Somehow, the dam just "didn't stay there."

It would become a commonplace that Mulholland was forever crushed by guilt over the collapse of the St. Francis Dam. George Bejar, the Chief's longtime driver, described some years later the toll the disaster had taken on his boss. "If I could only sleep at night," the weary Mulholland told Bejar as the inquest dragged on.

"It was that damned dam that killed him," said Bejar, whose recollections of Mulholland's later years range from the comic to the tender. "He never lost his sense of humor altogether," Bejar said, "but he did timid it down a little." Once, said Bejar, during a trip to Pendleton, Oregon, on some dam-consulting business, they stopped at a local restaurant. Informed by the waitress that the special of the day was "codfish balls," Mulholland nodded sagely. "That is the best part of the fish," he replied. Bejar also recalls being the only person the proud Chief would permit to give him a hand when his age-related palsy began to worsen. "He'd lean out over the rail to check something way down the face of a dam, shaking so bad I worried he'd go on over," Bejar said. "I'd grab his coattail and he'd give me a dirty look, but he let me do it. If a stranger tried to lend a hand, though, he'd yank right away."

Nephew William contended that Mulholland was not broken by the dam's failure, however. His uncle was indeed saddened, and he acted with great bravery and compassion when he asked that blame for the tragedy be fastened on him. But, William said, "He was not broken by that mishap because he never accepted the responsibility of something that was beyond his power."

Arguments persist as to whether Mulholland should have known better when he chose the St. Francis site. The well-received *Man-Made Disaster,* published in 1963 by Santa Clara Valley rancher and self-taught historian Charles Outland, hardly exonerated Mulholland, but in synthesizing the several reports on the disaster, Outland concluded that, in fact, the collapse had actually begun on the east side of the dam where the rock appeared more stable.

This view was elaborated upon by geologist and civil engineer J. David Rogers in a lengthy 1995 article published by the Southern California Historical Society. Rogers suggested that not only was Outland likely correct in pinpointing the actual locus of the dam's failure, but that owing to limitations of geological science of the time, Mulholland could not have appreciated the tenuous makeup of the substrata at the dam site. At the time of Rogers's writing, he said, it had become common knowledge among geologists that ancient landslides had once blocked the narrows of San Francisquito Canyon with insubstantial deposits that the creek had eventually worn through (much as the Owens River had once sawed its way down to China Lake). Modern engineers might better understand the nature of these ancient deposits, as well as the complicated physics of "uplift," and might be better prepared to predict that a concrete dam of the nature Mulholland built was likely to fail, Rogers wrote. But, Rogers concluded, Mulholland could not have fully comprehended these factors at the time, even though, ironically, the Chief at first theorized that "earthquakes" had been the cause of the disaster.

An article by Donald C. Jackson and Norris Hundley Jr., published in a 2004 issue of *California History*, took Rogers to task, however, for what the authors described as an attempt to exonerate Mulholland, castigating Rogers for his closing comment that "we should be so lucky as to have any men with just half his character, integrity, imagination and leadership today." Jackson and Hundley quoted Mulholland colleagues J. B. Lippincott and John Freeman (an original member of the Aqueduct Advisory Board and a consultant on the Panama Canal), who shared their own private concerns during the inquiry that Mulholland had proceeded imprudently in building the St. Francis Dam without sufficient outside consultation, especially given the suspect nature of the foundations. Jackson and Hundley say that, while Mulholland suffered death threats and lived with an armed guard at his house for some time after, they lament that "professional colleagues did not publicly pillory him."

It is the position of the authors—professors of history who have written extensively about dams built in the late nineteenth and early twentieth centuries—that in fact dam-building science was more advanced than Rogers suggests, and that if Mulholland was not aware of certain basic precepts of concrete dam building at the time, he should have been. In their opinion, it was likely conceit that prompted Mulholland to make peremptory decisions concerning the construction of the St. Francis Dam and that kept him from seeking the advice of others. Because of his sense of "privilege," say the authors, Mulholland proceeded on his own course. Their final pronouncement was, "William Mulholland bears responsibility for the St. Francis Dam disaster," something of a stern judgment coming seventy-five years after the event.

Much of the post-failure analysis makes for compelling reading indeed, but given the facts of Elwood Mead's original condemnation of the choice of the dam site, the coroner's pronouncement that the Public Service Commission erred in granting total oversight for the undertaking to one man, and Mulholland's public acceptance of responsibility, not much that is truly startling has come to light, though it is likely that given the scope of the disaster, debate on the matter will never end. In the contemporary context, the concept of "accident" has lost much of its cogency, and in today's litigious society, the pronouncement of "human error" being involved in some calamity seems a certain pretext for either bringing criminal charges, or, at the very least, a massive lawsuit. In the aftermath of the "I am not a crook!" era, the very thought of a public official stepping forward to accept personal responsibility for a catastrophic incident seems beyond imagining.

Despite the continuing controversy and whatever the truth of the statement that Mulholland was not broken by the collapse, the incident marked for all intents and purposes the end of the Chief's career, a tragic turn for a man who just a few years before had been listed with Thomas Edison and Orville Wright as being among the country's most respected members of the profession by the American Association of Engineers. On November 14, 1928, some seven months following the coroner's verdict, the *Los Angeles Times* carried the news that Mulholland had stepped down on the previous day, a decision that came in the wake of one of his last acts of significance—the order to lower the level of the Mulholland Dam in Weid Canyon by twenty-one feet, to less than half capacity, while tests of its soundness were conducted. (Though no faults were discovered, the reservoir would be drained in 1931 for retrofitting and strengthening, and not reopened until 1934.)

"It is no secret that the collapse of the St. Francis Dam hastened Mr. Mulholland's resignation," the *Times* story stated. "The tragedy, which was felt keenly by the 'Chief' as he is called at the water department, has aged him. After the first shock of the disaster, Mr. Mulholland plunged into the work of repairing the break in the city's water system," the report continued, "and has been actively on the job every day since last March 13."

With fifty years of service to the water department, and at seventy-three years of age, he would be remembered not only as the builder of the Los Angeles Aqueduct but of a water system that had grown in value from $100,000 to more than $100 million, the story said, with 3,200 miles of mains and 285,000 customers within a city of 440 square miles and 1 million citizens. Water commissioner R. P. Del Valle praised Mulholland for having designed and built the aqueduct, "the boldest and most spectacular engineering work ever undertaken by an American city." Del Valle also commended Mulholland's typical foresightedness in leading the fight for the Colorado River Project and went on to say that "No engineering achievement in this country within the past half-century has exceeded in difficulty or merit that of William Mulholland."

In December, shortly after Mulholland announced his retirement, the Swing-Johnson Act authorizing Boulder Dam was passed, opening the way for the development of the entire California Southland. In March 1929, Harvey Van Norman took over as head of the newly reorganized Department of Water and Power, the entity that survives to this day. Mulholland was asked to stay on the payroll as a consultant receiving a stipend of $500 a month, a post that he held for nearly seven years.

Catherine Mulholland has written gracefully of her grandfather's twilight years, noting that the once loquacious Chief seemed to grow quieter, even as he began to accompany his children and grandchildren on outings that he had never had the time to enjoy before. She remembers one trip to Palm Springs with her own father, Perry, Mulholland's eldest son, behind the wheel and "Grandpa" dozing in the backseat when they pulled up at the newly opened Desert Inn. When her father returned to the car to announce that rates at the imposing hostelry began at $10 per night, "Grandpa" Mulholland came awake in a trice. "Ten dollars?" he shouted. "Did you tell them we didn't want to buy the place?"

In the end, she recalls, the family stayed at a set of decidedly inferior tourist cabins nearby, and the next morning her father and Mulholland were still squabbling over some minor issue when her father jammed the car in reverse and backed over a water hydrant by the motel's parking lot. There was a crash and a sudden geyser of water into the desert air, and her father went sprinting toward the motel office while her grandfather climbed out of the car to gape helplessly at the scene.

"Only the excitement and confusion of the scene registered on me then," she wrote, "but now, years later, the memory of that old man who had been one of the world's leading hydraulic engineers standing helpless by a broken water pipe outside a tourist cabin in the desert strikes me as one of those consummately ironic moments when the gods play their Olympian jokes and laugh in heartless derision at us mere mortals."

In those after-years, her family each summer took an unprepossessing house on the beach at Santa Monica for a month, and her Aunt Rose, still living in the family home, would often bring Mulholland down to visit. He would sit before a bay window and gaze out at where she and the other grandchildren played, Catherine recalls, but, "When we children waved to him from the surf, he did not wave back."

In October 1934, Mulholland fell and broke an arm, and the following December, at seventy-nine, he suffered a major stroke that left him paralyzed and unable to swallow solid food. For six months, he battled on, confined to a hospital bed in his own home, tended to by Rose and a day nurse with the perfect surname of Ironsides. Finally, on July 22, 1935, after rousing for a moment to call out some orders that a ship's lookout might wish to relay to the forecastle, he closed his eyes and rested.

ONE SOURCE with whom I spoke as I researched Mulholland's life and works asked point-blank, "Just what side of the controversy surrounding this aqueduct do you intend to come down on?" A question to be expected, granted, though the person asking it was not quite as opinionated as another to whom I divulged the focus of the work. At the news that I was writing about the aqueduct that made Los Angeles possible, she, a New Yorker of the Woody Allen school, quipped, "Oh, it's a tragedy."

Both the question and the comment reminded me of another such project that I undertook some years ago, an account of Standard Oil cofounder Henry Flagler's building of a 153-mile railway project from Miami to Key West, destroyed by a hurricane in 1935. That project had not gone unrecorded by historians, and it was not without its own attendant controversies. Flagler was hauled into federal court to defend himself against charges that he was running a forced labor camp, charges that were thrown out before the trial began. One brutal winter shortly following the completion of the Oversea Railway, there was talk of a massive suit against Flagler by storm-beset Great Britons certain that Flagler's project had forever altered the path of the Gulf Stream.

But controversies have only one small part of my fascination with such tales wherein one of the most powerful men of an era undertakes a project that most consider impossible and overcomes all obstacles. The attendant and inevitable controversies, the coda of obliterating hurricane or dam collapse, are significant points, without question, but they dangle from the shape of the underlying whole just as individual ornaments depend from a Christmas tree. It is the magnitude, the daring and the import of the whole, that matters.

It was in that spirit, then, that this writer conceived of having a conversation with Robert Towne, the person who is likely as responsible as anyone for bringing an awareness of William Mulholland and the Owens Valley water into the modern consciousness. There may be a Mulholland Drive, a Mulholland Memorial Fountain, a Mulholland Dam, and even a Mulholland Middle School (est. 1963), but none of those has been the recipient of an Academy Award. Furthermore, a chance comment of Towne's—to the effect that the turn-of-the-century events referred to in *Chinatown* constituted a real-life story just as powerful as the fictitious one—had ever after resonated with this writer.

A synopsis of the film may be in order: in 1937 Los Angeles, tawdry private investigator Jake Gittes is hired by a woman impersonating the wife of Hollis Mulwray, the chief engineer of the Department of Water and Power, to investigate her suspicions that Mulwray is having an affair. Soon after Gittes takes incriminating photographs suggesting that she is correct, those photographs end up in the newspaper, causing a scandal, and weakening the authority of Mulwray, who opposes the building of a proposed new dam that farmers in the San Fernando Valley are clamoring for. Interestingly enough, Mulwray opposes the dam because a previous structure that he built on a similar site collapsed.

In short order, Mulwray's body is found in one of his own reservoirs, and Gittes suspects foul play. His investigation leads him to believe that Noah Cross, the former owner of the water company before it was sold to the city, is involved in a plot to create the illusion of a water shortage (water is dumped via sewers into the ocean at night, etc.) so that the new dam will be built, increasing the water supply and increasing the value of San Fernando Valley lands that Cross and his business partners have been buying up on the sly.

A featured subplot of the film involves Noah Cross's previous rape of his daughter (played by Faye Dunaway), an act that resulted in the birth of a child, now a teen, who is at once a daughter and a sister to her mother. By the time the film begins, the character played by Ms. Dunaway has married Hollis Mulwray.

During the course of his investigation, Jake Gittes (Jack Nicholson) falls in love with Mulwray's widow, and though he eventually confronts the all-powerful Cross with proof of his actions, it comes to nothing. Gittes is discredited and Mulwray's widow is killed in a shootout when she tries to flee with her daughter/sister. In the end, Jake is advised by one of his underlings, "Forget it Jake, it's Chinatown," an echo of advice Gittes himself delivers to a client early on in the film, "You dumb son of a bitch, you gotta be rich to kill somebody, anybody, and get away with it."

Even though the word "aqueduct" is scarcely mentioned in the film (and in that instance by Cross, who is explaining his own schemes to Gittes), such disinterested sources as the Internet Movie Data base assert, "The plot is based in part on real events that formed the California Water Wars, in which William Mulholland acted on behalf of Los Angeles interests to secure water rights in the Owens Valley," thus suggesting how powerful a film can be in reshaping the popular perception of actual events. But in this instance, the point of speaking with Towne was not so much to debate history as to find out what had drawn him, a filmmaker, a storyteller, and a native of Los Angeles, to the contours of a true story from the dimmer halls of history. Or to put it another way, what about this piece of actuality had drawn the interest of a man who could have simply "made it all up?"

It took some effort, but finally Towne agreed, and he began by confirming that he had indeed grown up on the Palos Verde Peninsula, in San Pedro, once a separate city near the mouth of the Los Angeles River. It was originally an unassuming port district, officially annexed by Los Angeles in 1909, and Towne is quick to point out that anyone who does not call it San *Pee*-dro marks himself as a rank outsider. Towne, born in 1934, attended Redondo High School and later graduated from Pomona College in Claremont, where he studied English and philosophy with little thought as to whether those were practical fields. "That's just what interested me," he says. Today, he lives in Rancho Palos Verdes, only a few miles from San Pedro as the crow flies, if light years away in terms of privilege.

When asked if he considered himself an Angeleno, given the setting of certain films he is often identified with—in addition to *Chinatown* and its more modestly received sequel, *Two Jakes* (focused on real estate and oil, made in 1990), he wrote and directed two other LA-based films, *Tequila Sunrise* (1988) and *Ask the Dust*

(2006)—Towne replied that he supposes he is, though he did not think much about such things when he was younger. As he was growing up, he thought of himself as a "South Bay" kid, a much more focused identity.

In any event, and despite the fact that he grew up on the banks of the very river whose vagaries of flow led to the city's search for alternate water sources at the turn of the twentieth century, Towne claims that he was unaware of the lore surrounding Mulholland's aqueduct and the controversy surrounding the city's acquisition of Owens Valley water rights until he was thirty-five and searching for a story of intrigue that could be melded into a 1930s Los Angeles setting.

At the time of the writing of *Chinatown,* Towne was still struggling. Despite having done uncredited work on *Bonnie & Clyde* and *The Godfather* (he would go on to write *Shampoo, Heaven Can Wait,* and many more), he was still a relative unknown. *Chinatown* would become his first original credited work, and though the film would have its fabled popular and critical success, including eleven Oscar nominations, the expectations of Paramount Studios at the time of production were modest. When asked whether he or anyone involved with the production ever expressed any concerns about his introducing actual historical material into a thriller, he laughs.

"No one was concerned," he says. "I don't think anyone I was dealing with thought Los Angeles even had a history."

As to the propriety of transposing events that had actually taken place thirty years or more before, Towne says that it did not concern him at all. He points out that violence over the aqueduct's appropriation of Owens Valley water continued into the late 1920s—and even to the late twentieth century.

In fact, a recent *Los Angeles Times* story carried the reminiscences of Mark Berry, a Lone Pine resident who spent thirty days in juvenile detention for dynamiting a section of the aqueduct on a September night in 1976. The following day, someone strapped a stick of dynamite to an arrow and shot it into the Mulholland Memorial fountain, though that charge failed to go off.

Similarly, one Owens Valley librarian volunteers that any number of her patrons who are rangers on public lands in the area regularly complain that vandals—or protestors, depending on one's perspective—persist in opening floodgates to dump Owens River water out of its channel short of the Los Angeles diversion point. Furthermore, while his story is obviously fictitious, Towne is unapologetic about his employment of facts: "I consider it historically accurate," he said, insisting that his script is "based upon the evidence that a group of influential men from Los Angeles profited greatly from insider knowledge that water would come to the San Fernando Valley."

When asked if he has ever heard the tale of the Los Angeles Department of Water and Power official attending the film's premiere and protesting that all had been fabricated with the exception of the "incest," Towne is amused. He has not heard that story, he claims, but he agrees that it is a good one. Asked where the subplot of Noah Cross's illicit relationship with his daughter actually came from, he answers without hesitation. "It seemed obvious to me that a man like Cross would rape indiscriminately. He would rape the Owens Valley. He would rape his own daughter. Whatever he wanted he took. The idea simply presented itself to me immediately."

Given that any Hollywood production necessarily aims to recoup its daunting costs by appealing to a broad audience, one could wonder what made Towne confident that a controversy nearly three-quarters of a century old, rooted in the convoluted issue of water rights, would strike a chord in the public consciousness. After all, even he has admitted that no major studio would want to deal with a story of *Chinatown*'s complexity today. "Well, it is a detective story," he says, "and I will grant that most detective stories deal with far more exotic matters than the ownership of water. But on the other hand, I found the very ubiquity of water in everyone's lives to be compelling. Everybody needs water. It just seemed to me that I could write a very persuasive story about it."

As to William Mulholland's portrayal in the film, and while Catherine Mulholland may have complained about the portrayal of Hollis Mulwray as an ineffectual cipher, it might seem to many that in his screenplay, Towne let the water superintendent off rather easily, compared to Noah Cross, for example. "I never thought William Mulholland was a bad guy," Towne agrees. "But the men whom I combined into Noah Cross for the film, they were monsters."

And as to who these monsters were in actuality? "One prototype was Harrison Gray Otis," Towne says, "and also Harry Chandler [Otis's son-in-law and ultimately a *Times* editor], and some others went into the character of Cross. They were all motivated by greed, pure and simple." Accordingly, that became the theme of *Chinatown,* he says: "The lust for money and power."

Regarding the role of Fred Eaton and Mulholland in acquiring the water and bringing an aqueduct to Los Angeles, Towne was asked to make a final verdict: In the end, was the building of the Los Angeles Aqueduct a great achievement? Or an example of American business skullduggery at its finest?

"I think it was a bit of both," Towne replied. "Mulholland's physical achievement was significant, but if he had been willing to build a dam further up the Owens Valley, ranchers and farmers there could have flourished along with Los Angeles."

As to how much Mulholland knew about the plans of Otis, Chandler, and others to profit from the sales of what would have been otherwise worthless lands in the San Fernando valley, Towne is less certain. "It is hard to say how much he knew," Towne says, "but he was a pragmatist. He wanted to bring water to Los Angeles, and if it took getting along with people like Otis to allow it to happen, he may have overlooked some things."

For the record, no one has ever put forward credible evidence of collusion between Mulholland and Otis. He had only spoken to Otis twice, Mulholland testified in 1911, once on official business and once when he ran into him in a store on Christmas Eve. The most likely explanation for the land syndicate's actions seems obvious at this remove: influential businessman Moses Sherman, who had become a member of the water commission by 1905, got wind of Mulholland's intentions concerning the aqueduct and its route and suggested to his syndicate partners that they quickly exercise their options on the San Fernando Valley lands.

In the end, Towne disclaimed any imaginings of the success his first screenplay would engender. When the lights went up following the film's first public screening, he told himself, "Well, maybe it's not a complete disaster." Of course he could not have predicted the acclaim and lasting influence of the film, Towne says, but there was one thing that did surprise him. "I never expected the controversy that it caused." Then he adds, with a dramatist's timing, "Of course I didn't mind it, either."

CITY OF ANGELS

IT WOULD NOT APPEAR THAT ROBERT TOWNE VIEWS HIMself as the final authority on William Mulholland and his legacy, but it is just as clear that he feels he has nothing to apologize for. And while the historical references alone cannot account for the ultimate power of an exquisitely made film, it seems just as impossible to imagine *Chinatown* working its legendary spell upon viewers without the allusions—however distinct from reality—to the actual events that took place.

As cultural commentator Ian S. Scott has recently noted, *Chinatown,* however fictionalized, retains its power for contemporary viewers because it "is in actual fact a prophetic vision of L.A. to come and a resemblance of the developments and personalities that have dominated recent times rather more than the Depression era." The film continues to engage, in Scott's view, "in the position of historical signifier for a series of developments that somehow delineate the identity and outlook of California in general, and Los Angeles in particular."

This writer might put it more simply: Academy Award–winning scripts are not written about dull matter. Whether true or fictive, a good story is a good story. And one thing is certain: without Towne's choice of subject matter upon which to base his narrative, this writer would likely never have extrapolated "Mulholland" away from "Drive," and the account at hand would likely not have come to be. Towne may in fact have been inspired to invest his fictive story with the power of history; the present aim is to demonstrate that history contains compelling drama. It is impossible to know what thoughts ran through the mind of William Mulholland as he sat in an overstuffed chair bundled in a suit before a bay window of a beach house at Santa Monica and stared out at the relentless surf, but one can suspect. "I took a vacation once," Mulholland once recalled. "I spent an afternoon at Long Beach. I was bored to death from loafing and came back to work next morning." Certainly, even if he was not afflicted at times by thoughts of the great losses from the St. Francis collapse, he must have felt a bit bereft.

For more than fifty years he had "only wanted the work," and now the work was gone. He had tended the flow of the water to the City of Angels from very nearly the first day he laid eyes upon it, had nurtured, grown, and protected that flow until it became a cascade that fed a city beyond anyone's capacity to imagine. Surely—even if history had decreed that instead of coming west, William Mulholland was to lose an injured leg and become a Cincinnati beggar—the city would have found its water somehow, somewhere, sometime. Surely. But on the other hand, Mulholland is the essential fact of the historical matter, and how could he have been content to sit quietly without his work to do?

In December 1932, his old friend Fred Eaton finally realized a three-decades-old ambition when he sold his Long Valley cattle ranch to the City of Los Angeles. But the price paid by the city—$650,000—was far less than the $1 million he had always dreamed of, and the money was not only too little but too late. In another of history's ironies, Eaton had been among the victims of the Watterson bank collapse, which took with it the $200,000 that Eaton had on deposit. In 1932, his ranch was foreclosed upon to satisfy a mortgage that had been arranged by the Wattersons without Eaton's knowledge. Thus, the city's purchase, at the ranch's appraised value, amounted to little more than a wash for Eaton, who by then had suffered a stroke and was in failing health.

In the aftermath of the sale and the settling of a mountain of debts, Eaton—then living in Los Angeles with his son Burdick "Bud" Eaton, chief engineer for the Huntington Electric Railroad (forerunner of today's Metro system)—turned to one last concern. He sent word to William Mulholland, to whom he had not spoken in nearly five years, that he would like to meet.

Hearing the request, Mulholland, who had never spoken an untoward word about Eaton in public, donned his hat without hesitation and had himself driven to the Eaton home. "Hello, Fred," Mulholland said, upon being ushered into Eaton's room. And in the few minutes that ensued, one very old wound was healed. (While no details of that final meeting were recorded, Hal Eaton, Fred's great-grandson, says that the story was likely passed along to writer Remi Nadeau by Bud Eaton and that in any case, Nadeau's version reflects what has been passed down within the Eaton family.)

On March 11, 1934, Eaton died, the news brought to Mulholland by his daughter Rose. Mulholland sat pensively for a few moments, then finally glanced up at her. "I've been dreaming about Fred," he told her. "Three nights in a row. The two of us were walking along, young and virile like we used to be." He paused again. "Yet I knew we were both dead."

A little more than a year later, both *were* dead. Mulholland's body lay in state at City Hall, and on July 26, 1935, his funeral services were held at the Little Church of the Flower at Forest Lawn Memorial Park. He was eulogized by attorney Joseph Scott, who called him the "human dynamo" who made Los Angeles what it had become. Mulholland's was a huge soul, Scott said, his example an estimable one, given "these days of alibi artists and buck passers and time servers." Another contemporary wrote, "He had little piety, much strength, great ambition. There is no one else in sight, past or present, whom Los Angeles is more likely to remember." His estate, estimated at $700,000 (composed principally of his home and 640 acres of ranch land assembled over time in the San Fernando Valley), was to be divided between his children—Perry, Tom, Lucile, Ruth, and Rose. Rose, in addition to her share, received the home near Third and Western.

As the saying goes, one could write a book about all that has devolved in the time since Mulholland's passing. In fact, many have been written, and it is likely that many more will be written. With Fred Eaton's ranch in hand, the long-discussed dam at Long Valley was built, though the dream of using those waters to feed a vast irrigation project in the Owens Valley never materialized. The city acquired water rights from the Mono Basin Watershed above the Owens Valley in the 1930s and 1940s, and a second aqueduct paralleling the first was completed in 1970 to bring more water from the distant region to the city.

By 1945, Los Angeles had acquired 88 percent of the total town property in the Owens Valley, paying a premium of anywhere between 45 and 120 percent above 1929 appraisal figures in lieu of reparations payments. By 1945 the city was also the owner of

278,055 acres of farmlands in the valley, just under 99 percent of the total. While some of that property has been sold off and returned through various exchange programs establishing reservations for Native American tribes and the like, it is estimated that the city still owns at least 25 percent of the valley floor, including more than 90 percent of the land that is usable. According to Los Angeles Department of Water and Power's (LADWP) Fred Barker, 94.4 percent of the entirety of Inyo County is in the hands of the US government or the State of California. The City of Los Angeles owns 3.9 percent. Private lands amount to 1.7 percent.

One study says that there are today about 3,000 irrigated acres planted in hay and alfalfa, and 8,000 acres are irrigated as pasture, with one private citizen claiming to make several thousand dollars a year raising chili peppers. Meanwhile, litigation between the city and Owens Valley officials over water issues has continued into the twenty-first century, including wrangles over the city's rate of groundwater pumping and the restoration of the lower Owens River and the marshlands surrounding Owens Lake, as well as a recent fight in Lone Pine to reclaim control of business frontage along US 395.

Rancor over issues first raised in 1905 persists, not only on the page and in the courtroom, but in ordinary life as well, as the tale told to the *Los Angeles Times* by latter-day saboteur Mark Berry attests: On the night of September 14, 1976, while he and pal Robert Howe were killing time, waiting for their girlfriends to get off work at a local ice-cream parlor, they bought a six-pack and drove to a turn-out close by the former course of the Owens River near Lone Pine. They were walking along the dry riverbed when Berry says his friend Howe boiled over. "They're not letting any water out. I'm going to fix this once and for all."

In short order, the pair had availed themselves of two cases of dynamite stored at a country trail-clearing facility. They drove to a set of gates on the aqueduct where they lodged one case in the middle of the structure, lit a five-foot fuse, and ran for cover. In minutes, there was an explosion, and soon 100 million gallons of aqueduct water were rushing back into the riverbed and toward Owens Lake. The two were later apprehended, with the older Howe serving ninety days in the county jail and high-schooler Berry spending a month in juvenile detention. Part of his sentence required that Berry enroll in a nearby junior college, where he studied—what else?—rocket science.

Getting caught and sentenced was the best thing that ever happened to him, Berry later said. He returned to Lone Pine in 2000 after a career in aviation engineering had taken him around the West. His new employer was, as irony would decree, the Los Angeles Department of Water and Power. As Berry told a somewhat stupefied reporter, "There was a time when the DWP did whatever it wanted to around here," Berry said. "But times have changed, and so have I. The DWP has done heroic work on behalf of the Owens Valley."

Berry rose from the lawn chair where he had been telling his tale and led the *Times* reporter to an overlook atop a cliff on the outskirts of Lone Pine. He gestured out at a vista of mountainside streams, lava beds, and plunging rock formations, the likes of which have drawn any number of filmmakers, artists, rock hounds, recluses, fishermen, and tourists of various stripes. Berry was happy, for one, that urban sprawl had been precluded in the Owens Valley. For him, he said, it was "the most beautiful place on Earth."

For its part, the DWP, though surprised to learn that a former aqueduct bomber was now on its payroll, seemed inclined to let bygones be bygones. The incident was "a reflection of a period of much more tense relations between people in the Owens Valley and

L.A.," said aqueduct manager Jim Yannotta, who pointed out that Berry had also been held accountable and served his time. "We are grateful that relations have improved."

As for who could have fired that stick of dynamite into the Mulholland Memorial Fountain, the day after his own wrong-headed action, Berry claimed not to have a clue. "I don't know who fired it," he told the reporter with a smile. "But it wasn't me."

One would likely be cautious about raising a toast to the memory of Bill Mulholland in any Owens Valley tavern where locals congregate, just as one might want to avoid praising Andrew Carnegie in any mill town bar near Pittsburgh, where memories of a steel strike quashed more than a century ago by the fabled entrepreneur could still prompt a fistfight. History is not a dead issue in certain places.

Catherine Mulholland says that for many years her family regularly signed into hotels in the Owens Valley using her mother's maiden name. And even though the August 31, 1990, issue of *Life* magazine listed Mulholland among its selection of the 100 Most Influential Americans of the Twentieth Century, along with Albert Einstein, the Wright Brothers, and Henry Ford, the honor did not necessarily cut any ice in some quarters in Los Angeles. A series of *L.A. Weekly* articles published shortly thereafter characterized Mulholland not as a hero, but as a "Vengeful Master Builder," he of "hooded, flinty eyes and grim and vindictive mouth." Truly, "an entire valley had been destroyed by the Self-Made Man's personal feud," the piece assured readers, before concluding, in neobiblical terms, "its history run through the shredder, its people mocked and terrorized and cheated, and their farms spread with salt."

Then again, there are such publications as that by University of California at Santa Barbara economist Gary Libecap, *Owens Valley Revisited*, which argues that the Edenic view of the preaqueduct valley is little but a myth. Farming in the region was always an iffy business, Libecap points out, making reference to many a fact and figure, and most of those to whom offers were made were happy to take the city's money and decamp. It is Libecap's conclusion that many residents used the oft-cited characterizations of themselves as hapless victims in order to secure lucrative offers for land they would have had trouble selling otherwise.

In November 2013, any number of events celebrated the centennial of the aqueduct water's arrival in the San Fernando Valley, including the guiding of a hundred-mule pack train all the way from Independence that culminated in a clopping parade through the streets of Glendale. At the Cascade below Newhall Pass, DWP officials loosed a torrent down the spillway at 1:15 P.M. on November 5, 2013, one hundred years to the minute after Mulholland had presided over the ceremony celebrating the water's original arrival. According to the *Times,* an actor playing Theodore Roosevelt (who did not attend in 1913) reminded those present that he had approved the project "for the good of all," and Los Angeles Mayor Eric Garcetti put a contemporary spin on Mulholland's fabled line when he pointed to the water rushing down the concrete steps and said to the crowd, "There it is. Conserve it."

Coming as it did in the midst of another cycle of drought in California, observation of the centennial was also the spur for any number of renewed criticisms of the Department of Water and Power, including a rehash of the original land acquisition controversies, the "Chinatown" conspiracies, and a thoroughgoing litany of misdirected contemporary water practices and waste. One could stand atop Mulholland Drive, "the supreme vantage point for the entirety of Los Angeles and the San Fernando Valley," and gaze out upon what has come to be, "and feel like Christ—or the Devil," one commentator said. "Mulholland Drive allows both roles."

According to DWP figures, from 1913 to 1990 the Los Angeles Aqueduct supplied on average about 75 percent of the city's water, with as much as 532,000 acre-feet of Owens Valley water flowing through the twin conduits in 1969, more than 80 percent of total consumption for the year. In normal years, rain and snow melt brings anywhere from 350,000 to 725,000 acre-feet into the valley. As much as 100,000 acre-feet is lost to evaporation and drainage into the ground, with another 80,000 to 100,000 acre-feet going to irrigation and stock watering in the valley. Another 5,000 to 6,000 goes to Native American lands, and 8,000 to 9,000 is used for recreation and wildlife purposes. The remainder goes into the aqueduct, though owing largely to the 1994 court-ordered restorations to Mono Lake, the Lower Owens River Restoration Project, and an air quality mitigation settlement requiring seasonal flooding of the flats surrounding Owens Lake, the net flow of the aqueduct has been reduced to between 100,000 and 150,000 acre-feet on average, or somewhere between 30 and 40 percent of the total city supply. The city still draws about 10 percent of its water from local groundwater wells (i.e., the Los Angeles River watershed), with the remainder purchased from the Metropolitan Water District, which is supplied by both the Colorado River Aqueduct and the California Aqueduct, which originates in the Sacramento–San Joaquin River Delta east of San Francisco.

The effect of the current drought cycle, with Sierra Nevada snowfall less than 40 percent of what is normal, on the "limitless" Owens Valley Watershed is reflected in the following DWP statistics: from July 1, 2011, to June 30, 2012, the Los Angeles Aqueduct supplied 266,700 acre-feet of water, or 49 percent of the city's supply; in 2012–2013, that figure dropped to 113,400 acre-feet, or 20 percent of what the city consumed; in 2013–2014, the aqueduct carried only 61,000 acre-feet, about 10 percent of the 580,000 acre-feet used.

During a February 14, 2014, appearance in drought-stricken Fresno, President Obama called upon an end to battles between the interests of Northern and Southern California and between agricultural interests and those of the cities. However, even he steered well clear of taking any specific sides in any water dispute for, as he said, he "wanted to get out of California alive."

Though some today lament the fact that no compromise was reached that could have somehow served the interests of both the Owens Valley and the City of Los Angeles, that view disregards any number of realities of the time. And it might also be noted that hindsight is always amazingly keen.

Certainly, if there had been a Reclamation Service project undertaken in the Owens Valley and if all the waters of the Owens River had been diverted to that use in 1905, there would be a great deal more alfalfa being grown north of Big Pine today, and there would likely be a fair amount of vegetable farming and orchard tending going on in the valley as well. In an 1892 interview with a Riverside newspaper, Fred Eaton presented a dazzling view of the agricultural possibilities: "There is an orange orchard doing well in the lower valley," he said, "and tomatoes ripen through the winter. I found all of the varieties of deciduous fruits, berries and grapes. . . . Cherries are also successfully grown. Apples, raisins, prunes and almonds excel, and I ate peaches of unusually fine flavor."

Had that reclamation project trumped Mulholland's, there could well be three or four times as many residents in the Owens Valley as are there at present, or even more, along with a corresponding greater number of schools, housing developments, and big-box retailers. But it is equally easy to argue that had William Mulholland not fought for and built the Los Angeles Aqueduct, Southern California as we now know it would not exist.

In the end, it seems that only one thing is inarguable: conflicting passions that swirl in the wake of William Mulholland and the building of the aqueduct will never be resolved. Given the ultimate scope of his achievement and the outsize nature of his personality, the fuel of controversy is inexhaustible. Mulholland himself must have surmised something of the complexity his legacy would entail, though it is an open question just how much he would have agonized about it.

Unquestionably, Mulholland was no ordinary civil servant, and as engineers go, he scarcely represents the norm. He took on an impossible project and brought it to fruition without the lure of a penny beyond his stipulated pay, largely because, as one writer put it, "to his simple and rugged heart, a great Los Angeles meant a big Los Angeles." What he sought was work that mattered, and those seeking to understand what drove him might consider a statement he once made to a reporter looking to learn what made Mulholland tick. "Damn a man who doesn't read books," he said. "The test of a man is his knowledge of humanity, of the politics of human life, his comprehension of the things that move men." Uncharacteristic sentiments for a civil engineer perhaps, but therein may lie his secret.

In that brief autobiography sketch written in February 1930, not long after his retirement as chief, Mulholland spoke proudly of his accomplishments. "We began with three or four men, and we now employ some three thousand," he said, still identifying himself as an employee of the water department. "Daily I receive letters from water works men all over the country inquiring how things are done here and how we keep up with the ever increasing rapid expansion," he said, before adding with no apparent irony, "I seem to have more of a reputation outside of the City of Los Angeles."

In defense of his reputation, Mulholland had a few things to say: "No politics ever got into this office. I have had to be rather firm in dismissing councilmen seeking positions for friends or henchmen," he declared. "I had no personal interest in politics but I had good staunch friends who, knowing that my sole object in life was to make things go right in this Department, devoted themselves to keeping things off my back." (A coworker recalls Mulholland's response to one brick maker who appealed endlessly to have his materials purchased for use on the aqueduct: "After mature thought on the matter," Mulholland finally wrote the brick maker, he had decided to build the aqueduct "of leather.")

In sum, it seems he felt there was little to apologize for. "No other single man in a single life time has built such a water department as we now have here in the City of Los Angeles. I have served the city for 52 years, and am the oldest in service in the State of California." So far as he was concerned, his life was an open book. "I belong to several professional organizations," he said, "but I never have been a member of any secret order. Although never a soldier I have devoted most of my life to the service of the City of Los Angeles, the County of Los Angeles and the State of California." The words were the definitive close for consulting engineer William Mulholland.

In that characteristically terse summation of his life's work, there was only one reference to the building of the aqueduct: "After a few years of drouth which began in 1895 the city was compelled to go to the Owens River in Inyo County, California for its water." Not long after he had delivered this model of understatement, Mulholland agreed to lead a visitor about the area near where Mulholland Memorial Fountain now stands, with its plaque reminding visitors of the foresight, vision, and engineering ability of the man who "made possible the rapid growth and industrial development of this community."

Mulholland had in tow his first biographer, the young USC graduate student Elizabeth Spriggs. Not far from the site of the shack where he lived when he received his promotion to foreman of the ditch-tending gang, Mulholland stopped the history student to point out a two-foot-thick live oak growing near the intersection of Riverside Drive and Los Feliz Boulevard.

It was the morning of March 22, 1930, when they took their stroll, Ms. Spriggs recalled, but Mulholland began to describe with remarkable clarity a day some fifty years before, when he was shoveling out the channel of a ditch that once ran where they were now standing. That oak tree was a three-inch sapling at the time, Mulholland told her, and had been about to topple into the water where he was making his cut.

Mulholland described stopping to take up the sapling. He climbed out of the ditch and walked a few steps to a spot he thought would make a suitable spot for a tree to grow. He replanted the sapling there and tended it for as long as he lived nearby. And a half-century later, there it was still, he pointed out, now a towering tree. Mulholland pressed his palm against the oak's thick trunk, Spriggs says, and gave her an inquisitive look.

"I saved its life once," he told her. "I wonder if it is conscious of my presence today."

He was speaking of a tree, but his words might as easily have been directed at a place.

NOTES

As this is not a work of traditional scholarship, and in the interest of avoiding distraction for the general reader, I have dispensed with the use of footnotes in the text itself. Much of the story is drawn from contemporary newspaper accounts and LADWP documents, though these materials have often been previously reported piecemeal and in varying contexts, according to the overriding thesis of an individual volume. The attempt here is not one of historical archaeology, though certain heretofore unexamined materials play their part. Nor is there any political agenda intended. The goal is to trace a factual story of remarkable, other-era accomplishment with a somewhat larger than life individual as protagonist. Still, in order to aid those interested in further reading, I have endeavored to give appropriate credit to sources, either in the text, or, where singular contributions seem to merit it, in the following notes. Any oversight in this regard is unintentional.

AUTHOR'S NOTE

The original Mulholland Drive ran from Calabasas to Cahuenga Pass near the Hollywood Bowl (a kind of dogleg off today's US 101 through the Santa Monica Mountains) and was actually christened "Mulholland Highway" in December 1924. The most popular leg of today's Mulholland Drive is the 10.5-mile stretch between I-405 and the Hollywood Freeway. Completed later was a 35-mile section westward from Calabasas, crossing Topanga Canyon and terminating at the Pacific Coast Highway at Leo Carillo State Park. It is known today as Mulholland Highway. The entire route is sometimes referred to as the Mulholland Scenic Corridor, though one section between Calabasas and the 405 is an unpaved fire trail open only to hikers and bikers.

The material regarding Mulholland Drive as a trysting place is drawn from the introduction to Catherine Mulholland's original manuscript, "William Mulholland and the Making of Los Angeles."

Ms. Mulholland's comments regarding previous biographies of her grandfather are found in the preface to her original manuscript.

Ms. Mulholland's remarks on *Chinatown* and other personal experiences as a Mulholland are found in the introduction to her original manuscript. The much-reduced published version of the original manuscript is entitled *William Mulholland and the Rise of Los Angeles,* hereafter referred to as *WM&RLA*.

In *WM&RLA,* Ms. Mulholland cites a "lead article" in the *New York Times* of May 1, 1991, in regard to *Chinatown* as documentary. This author finds two *New York Times* articles, both by Robert Reinhold, that contain nearly identical references to *Chinatown* as a virtual documentary chronicling the alleged nefarious practices of Los Angeles officials. One piece on April 23, 1991, concerns the water-supply difficulties facing Las Vegas; the other, on May 18, 1991, concerns a settlement pertaining to Mono Lake with Owens Valley officials.

This writer is greatly indebted to Christine Mulholland, niece of Catherine, for her insight into her aunt's devotion to chronicling the Mulholland legacy. Christine Mulholland's brother, Tom, a San Joaquin Valley rancher, is the lone surviving male bearing the family name.

An appreciation of Mr. Kaplan's career is included in an obituary published by the *New York Times* on March 4, 2014.

1. HOW DREAMS MIGHT END

Hopewell's account is found in "Transcript of Testimony and Verdict of the Coroner's Jury in the Inquest over Victims of St. Francis Dam Disaster."

Details of the dam's collapse are drawn principally from contemporary issues of the *Los Angeles Times,* the *New York Times,* and the *Los Angeles Herald.* Outland's *Man-Made Disaster* and *The St. Francis Dam Disaster,* edited by Doyce Nunis, are illuminating book-length treatments.

The report of the workers' lunch atop the St. Francis Dam the day of the collapse is from an interview with DWP engineer Clark Keely, in Matson, *William Mulholland,* page 54.

Mulholland's comments are drawn from contemporary news accounts and from the transcript of the coroner's inquest.

2. DISTANCE BETWEEN TWO POINTS

Details of the modern-day motor route tracing the path of the Los Angeles Aqueduct are drawn largely from the author's notes from a trip made in January 2013.

LADWP spokesman Fred Barker estimates that Owens Lake regeneration efforts have reduced the dust pollution by about 90 percent over past decades.

Mulholland's sobriquet for the river is noted in Layne, "Water and Power for a Great City," page 7. Layne's work, informative though scarcely to be found, was commissioned by the Department of Water and Power at the time of its fiftieth anniversary.

The trip undertaken by Fred Eaton and William Mulholland to the Owens Valley in late 1904 is originally referenced in multiple news accounts published in Los Angeles newspapers in 1905, attendant to public announcement of the Los Angeles Aqueduct project. It has been embroidered upon in countless retellings since.

3. LUCK OF THE IRISH

Mulholland's claim regarding his vote for Tilden comes from his "Sketch," page 1. His oft-repeated quote regarding the ease with which he made $25 walking across Panama also derives from the "Sketch."

DWP engineer and unofficial department historian Fred Barker once ran across information that suggested that the original San Pedro grantee never had children—it was likely his *brother* whose grandchild was Manuel Dominguez, the man who offered William Mulholland his first job in "water." When Fred Barker informed Catherine Mulholland that she might have made a mistake, she was unfazed. "That's hardly the only mistake in that book." Given that her original manuscript totals more than 1,000 pages and was planned as a two-volume study by Ms. Mulholland, one might be inclined to forgive the slip.

4. IN MYSTERY IS THE SOURCE

Census figures throughout—except for those cited by individuals in news stories, reports, and the like—are from the US Bureau of the Census.

The stories are recounted by Lippincott in "William Mulholland—

Engineer, Pioneer, Raconteur," his two-part appreciation of his former supervisor.

The oft-repeated anecdote regarding the incident that determined Mulholland's choice of career appears originally in Prosser, "The Maker of Los Angeles," page 43.

The often cited recollection from Mulholland's former roommate Brooks is contained in the LADWP Historical Records files: Ephemera, Box II:8.

The familiar riposte concerning Mulholland's qualifications comes from an interview with Allen Kelly, published in the *Los Angeles Times* on July 7, 1907.

The recollections regarding the Zanja Madre are found in Mulholland's "Sketch."

One can have a look at the original water main by peering over a balcony into the basement of the Avila Adobe in the Los Angeles Plaza Historic District. Americans with Disabilities Act requirements have resulted in the closing of the staircase leading down to the main itself, but a visit to this site gives some sense of what the city must have seemed like during Mulholland's early days in "water."

The story of Brooks's stepping aside for Mulholland is from Layne, "Water and Power for a Great City," page 54.

The *Times* account of the threat to the city's water supply and Mulholland's stalwart actions was published on July 16, 1890.

The material concerning Mulholland's wife and his mother-in-law, Frank, as well as other family matters, are drawn from Catherine Mulholland's original manuscript, "William Mulholland and the Making of Los Angeles," chapter 13, "Marriage."

The ground floor of today's DWP offices in the Ferraro Building on North Hope Street in downtown Los Angeles (surrounded by—not unexpectedly—an expansive water feature and fountains) contains a sizable exhibit on matters Mulhollandian. One of the more interesting items is an imposing longcase or "grandfather" clock that Mulholland gave his wife, Lillie, on their tenth anniversary. LADWP's Fred Barker makes a weekly descent from his office to wind the device, which bears the discernible imprint of the Chief's thumbnail from his own years of tending to the clock.

5. WHOSE WATER IS IT ANYWAY?

A clear picture of Los Angeles in the 1890s may be drawn from a glance at the pages of the newspapers of the day, accessible online and in various Southern California library holdings.

Intricacies of the bargaining over the city's acquisition of the water company were reported in news accounts of the day; the matter has been carefully detailed in Kahrl, *Water and Power;* Hoffman, *Vision or Villainy;* Catherine Mulholland, *WM&RLA;* Nadeau, *The Water Seekers;* and Ostrom, *Water & Politics,* among others.

Mulholland's observation on office detail is from McCarthy, "Water," December 18, 1937, page 28.

The "mighty memory" anecdote that is often retold may be in some ways apocryphal, says LADWP spokesman Fred Barker—detailed records for the location and size of mains, hydrants, and so on, go back well into the 1870s. On the other hand, and given his disdain for the sort of inquiry being made of him, Mulholland might well have enjoyed sketching out the information for bureaucrats on the spot.

The tale concerning Mulholland's advice to the water company's attorney comes from part II of Lippincott's "William Mulholland—Engineer, Pioneer, Raconteur."

6. A CIVIL SERVANT BORN

The information concerning the family's view of Fred Eaton's political motivations comes from a 2014 interview with Harold "Hal" Eaton, Fred Eaton's great-grandson.

For Eaton on the source of underground water, see the *Los Angeles Times,* March 23, 1900.

For Mulholland's visit to the summer home of I. W. Hellman, see Spriggs, "The History of the Domestic Water Supply of Los Angeles," pages 57–58.

Mulholland's public comments on the value of the private water company may be found in Mulholland, Office Files, Speech to the Pasadena Board of Trade, April 6, 1904; Speech to the Sunset Club, April 1905.

For public perception and passage of the water company acquisition bond issue, see the *Los Angeles Times,* August 11, 1901, and August 29, 1901.

On city attorney Mathews and the role of the Eastern bond markets, see Tzeng, "Eastern Promises," pages 42–43.

Mulholland's comments about his reliance on Mathews and his status among of the chattels of the water company have been passed along by many chroniclers, including Kahrl, *Water and Power,* page 23; Catherine Mulholland, *WM&RLA*, page 90.

Information on early water meter matters was passed along by Deborah Warner, chair of the Division of Medicine and Science at the Smithsonian's National Museum of American History. Those interested in the subject might begin by browsing the Smithsonian's website.

The location of the city's first water-meter installation is from Layne, "Water and Power for a Great City," page 55. Three years after he had stepped aside for Mulholland, Brooks was in 1890 promoted to assistant superintendent of the Los Angeles City Water Company.

A survey of Mulholland's office files reveals that, contrary to some contentions, Mulholland had been concerned with the cycles of drought in Los Angeles for a number of years. See Mulholland office files, DWP 04-22.5.

A readable summary of Mulholland's early struggles to balance supply and burgeoning population growth is in Moody, "Los Angeles and the Owens River," 1905.

Mulholland's statistics on the receding well and water table measurements are included in his "Fourth Annual Report" to commissioners.

The first public mention of the sewer dumping issue may have been in a *Los Angeles Herald* story on August 6, 1904.

7. ROAD TRIP

Harold "Hal" Eaton says that part of his great-grandfather's lifelong frustration with the City of Los Angeles stemmed from the fact that he had been on the verge of putting together a deal that would have enriched him considerably and benefitted ranching and farming interests in the Owens Valley as well.

Perusal of the "Commemorative Album" in the DWP Office Files provides a vivid picture of the nature of the 5,000-person Owens Valley at the time.

Campbell's recollection of the early trip to the Owens Valley is included in Layne, "Water and Power for a Great City," page 99. Early involvement of the Eaton family in the region is summarized in McGroarty's *History of Los Angeles County*, pages 462–463.

Eaton's early interest in Owens Valley water received considerable notice following the city's announcement—a piece in the *Los Angeles Express*, August 4, 1905, is typical. But there had been passing notice of his activities in the early 1890s, as the *Herald* story and a longer interview from the *Riverside Daily Press*, July 7, 1892, attest.

Chalfant published two editions of *The Story of Inyo*, the first of which (1922) made little mention of matters pertaining to the Los Angeles Aqueduct. The 1933 edition, however, is expanded by a seven-chapter, seventy-four-page-long coda that details what the author characterizes as a series of "merciless" predatory acts taken by the city and its various representatives—including Eaton—against the citizens of the Owens Valley. "Betrayal of Owens Valley" and "Unceasing Menace" are chapter headings that suggest the nature of the material.

Mulholland's summary of Fred Eaton's championship of the Owens River is from the *Los Angeles Times*, July 29, 1905.

Mulholland's testimony was delivered before the Aqueduct Investigation Board convened in 1912, shortly before the project's completion.

The characterization of various proposals to bring Owens Valley water to Los Angeles as wildly impractical was made in a 1905 letter by Arthur P. Davis to Secretary of the Interior Ethan Hitchcock and is quoted by Kahrl, *Water and Power*, page 47.

The intricacies and controversies surrounding the city's acquisition of water rights in the Owens Valley, first featured at length in Chalfant (*The Story of Inyo*), have formed the core of various more exhaustive studies, including those of Kahrl (*Water and Power*) and Hoffman (*Vision or Villainy*).

Mulholland's poetic evocation of the desert was penned as part of a report on the possibilities of Colorado River water for the Southland in the 1920s and is quoted by Ostrom, *Water & Politics*, page 3.

Mulholland's managerial style is described by Lippincott in "William Mulholland—Engineer, Pioneer, Raconteur," Part I, page 107.

8. DOWNHILL ALL THE WAY

The *Los Angeles Herald* on August 16, 1893, carried a report of Eaton's return from a trip through Inyo County during which he had made "the ascent of Mt. Whitney." Hal Eaton, great-grandson of Fred, reports that Eaton's youngest daughter, Dorothy, always claimed to have been conceived on top of Mount Whitney "but from calculation it would have been very cold in 1894 for that to have happened." While no permanent trail was established to the mountain's summit until 1904, it had been climbed as early as 1873, and there are accounts that several women first accomplished the feat in 1878.

In October 1893, Eaton served as Inyo County's representative to the Second Annual Irrigation Congress in Los Angeles. His father, Benjamin, represented Pasadena. See Sklar, "The Man Who Built Los Angeles," page 6, and the *Los Angeles Herald,* October 11, 1893.

As to the preliminary survey for an aqueduct "conducted at his own expense," Hal Eaton suggests that a *Los Angeles Herald* story on July 27, 1902, might have been a cover for travels Eaton had actually undertaken to help with that survey. The piece, headed "Fred Eaton as a Miner," reported that former mayor Eaton, "got home yesterday from a three-months' sojourn in Death Valley, the hottest spot in all the Mojave Desert. Mr. Eaton says he did assay work on twenty-five quartz claims he has located in the Mojave sink, and that he is highly pleased with the results of this work. As soon as the weather cools a trifle with the approach of the winter season, he will put gangs of men at work on his new-found claims. The ex-mayor says the heat was terrific—120 degrees most of the time. He is burned to the color of copper, and looks more like a Zunl priest than a former chief magistrate of the City of Angels." Hal Eaton suggests that the idea of pursuing quartz or gold claims in Death Valley during the summer is unlikely and that his great-grandfather may have actually been out in the Mojave assisting with that preliminary aqueduct survey, no copy of which seems to survive.

Eaton's lament regarding his decision to step aside in favor of the city's interests was made to a reporter for the *Los Angeles Express* on August 4, 1905. He provided other details in interviews with the *Herald* (August 5, 1905) and the *Los Angeles Times* (August 30, 1905).

9. REMOVE EVERY SPECTER

Mulholland's quote is from an interview published in the *Los Angeles Examiner* on May 29, 1920. The most fulsome personal account of his activities on the project is to be found in his 1916 Complete Report on

Construction of the Los Angeles Aqueduct, To the Los Angeles Board of Public Service Commissioners, hereafter referred to as "Complete Report."

Karhl and others make much of the ever-escalating figures (see *Water and Power,* page 87).

Details of the city's arrangement with Eaton are given in Mulholland's "First Annual Report."

For the story of Charley's Butte, see Chalfant, *The Story of Inyo,* pages 180–181.

A balanced reassessment of Lippincott's activities is provided by Hoffman in "Joseph Barlow Lippincott and the Owens Valley Controversy."

In addition to interviews with the *Times* and the *Express* mentioned previously, Eaton also spoke at some length with the *Examiner* on July 30, 1905.

Mulholland's interview with the *Times* regarding his decision to undertake the project appeared on June 2, 1907.

Mulholland's description of the possibilities of the Owens River resources and his description of the detailed proposed contours of the project is found in his "First Annual Report."

The author is indebted to LADWP's Fred Barker for a thorough clarification of "siphon" terminology and practice.

Though some have blamed Mulholland for failure to construct the dam at Long Key as part of the initial undertaking, even Chalfant points out that the decision was forced upon him by the consulting panel (page 340).

Representative Smith's efforts are described in Chalfant, *The Story of Inyo,* pages 354–355.

Mulholland's trip was detailed by the *Times* on June 23, 1906.

10. HAVE WATER OR QUIT GROWING

The account of Mulholland's triumphant appearance is from the *Los Angeles Herald,* August 16, 1906.

Information on the history of water treatment in the United States comes from documentation provided by the Environmental Protection Agency.

The quote from Henry Flagler is in Standiford, *Last Train to Paradise,* page 88.

Mulholland's brick-by-brick quote is from the *Examiner,* August 16, 1906.

The "primer" is quoted at length by W. S. B. in "Record of the Owens River Project." In her published book, Catherine Mulholland quotes a portion of the parody published in the *Evening News* on June 8, 1907 (*WM&RLA,* page 151).

11. BRICK UPON BRICK

The quote, as well as the description of early project activity, is from W. S. B., "Record of the Owens River Project."

A comprehensive treatment of the role of the bond market in financing the aqueduct project is found in Tzeng, "Eastern Promises."

A discussion of the impact of the Civil Service Commission upon the project is found in Mulholland's "Complete Report," page 252.

The account of Chaffee's survey trip is from the *Los Angeles Times,* April 10, 1908.

The anecdote involving Mulholland in the Oldsmobile is from Lippincott, "William Mulholland—Engineer, Pioneer, Raconteur," Part I, page 107.

Desmond's age is noted in McCarthy, "Water," January 1, 1938, page 4.

Taylor's notes and recollections would become the stuff of a memoir that he published in limited form in 1953. In 1982, University of Southern California historian Doyce Nunis published a readable and richly annotated edition, *Men, Medicine, and Water,* with the assistance of the Los Angeles County Museum of Art and the DWP.

12. FIRST SPADE

The private contractor's dour assessment of the terrain is from W. S. B., "Record of the Owens River Project," page 273.

A thoroughgoing assessment of work camp conditions is found in Van Buren, "Struggling with Class Relations."

Mulholland's rundown on work in the Jawbone and elsewhere comes from an interview with the *Los Angeles Times,* August 7, 1908.

Mulholland's appearance before the Chamber of Commerce is described by Heinly, "Carrying Water through a Desert," page 582.

Mulholland's declarations regarding the project's certain outcome are from W. S. B., "Record of the Owens River Project," pages 274–275.

13. BEST YEAR TO DATE

Mulholland's pork bristle quote, along with a summation of work to date, is found in an interview with Allen Kelly published in the *Los Angeles Times* on September 12, 1909.

The observation on the usefulness of the derby as a safety device, along with other detail, comes from Cross, "My Days on the Jawbone," pages 6 and passim.

The workforce census is from the *Times,* March 21, 1909.

The line describing the transitory nature of laboring crews is from Nadeau, *The Water Seekers,* page 50.

Widney's recollections are found in "We Build a Railroad."

Mulholland's explanation of siphon technique comes from his "Complete Report," pages 192–235.

Mulholland's personal interest in "hayburners" is noted in his "Complete Report" as well as in numerous incidences in his miscellaneous office correspondence. From a letter of July 12, 1907, to the Board of Public Works: "The Aqueduct is badly in need of four more horses for the equipment of an engineering party to survey the Francisquito Canyon portion of the line. I have found four horses admirably suited to the work, which can be bought for $900."

The comments from Smith are contained in the Kelly interview, *Times,* September 9, 1909.

14. FAIR MONETARY RECOGNITION

Smith's salary is noted in documents included in the "New York City Watershed Retrospective."

The editorial favoring Mulholland's raise is from the *Los Angeles Times,* October 19, 1909.

Mulholland's summation of the progress to date is from the *Los Angeles Herald,* November 4, 1909.

The Hansen anecdote is related by Lippincott in "William Mulholland—Engineer, Pioneer, Raconteur," Part II, page 163.

Mulholland's appeal to the Council and Board was reported by the *Herald,* December 2, 1909.

The layoffs were widely reported in Los Angeles papers, including the *Herald,* June 8, 1910.

Carnegie's quotes are from the *Herald,* March 23, 1910.

Mulholland was not shy about sharing his suspicion of the bankers' motives, repeating them in his "Complete Report," page 267.

The Heinly piece is "Carrying Water through a Desert."

Details of the inspection tour are from the *Herald,* November 4, 1910, and the *Los Angeles Times,* November 6, 1910.

The strike was first reported by the *Herald* on November 4, 1910.

The pay scales are detailed by Van Bueren, "Struggling with Class Relations at a Los Angeles Aqueduct Construction Camp," page 30.

News of the bond issue and Mayor Alexander's reservations were widely reported by local papers, including the *Herald,* November 11 and 22, 1910.

15. FITS AND STARTS

The detail of Taylor's innovative measures is found in his *Men, Medicine & Water,* pages 113–114.

The *Los Angeles Herald* carried word of Mulholland's view of tufa on November 27, 1910. The LADWP's Fred Barker suggests that time has proved the quality of the tufa-stretched concrete to be spotty, with a fair amount of that material having been replaced in recent decades.

The tale of Mulholland's testimony is from Lippincott, "William Mulholland—Engineer, Pioneer, Raconteur," Part I, page 107.

News of the arrangement with the Marion concern was carried by the *Herald* on November 10, 1910.

The strike was noted in both the *Los Angeles Herald,* February 10, 1911, and the *Los Angeles Record,* February 16, 1911.

Details of the Elizabeth Tunnel accomplishment are from Mulholland, "Sixth Annual Report," pages 38–41.

Details of the McNamara defense are included in Darrow, *The Story of My Life,* pages 184 and passim.

16. FALLOUT

A discussion of the impact of the Homestead Steel Strike on labor is to be found in Standiford, *Meet You in Hell,* page 233 and passim.

For Steffens's perspective on the matter, see his *Autobiography,* page 683 and passim.

Mulholland's speech before the women's club was covered by the *Los Angeles Times* on November 28, 1911, as was his follow-up before the men's club the following week, on December 3, 1911. A copy of the speech is in Mulholland's Office Files, WP04-22:24.

Mulholland's letter to the board is in his Office Files.

The troubling news from the Kountze Brothers was reported by the *Times* on January 12, 1912.

The headlines are from the *Times*, February 11, 1912; see also Tzeng, "Eastern Promises."

The comments of Mulholland and Chaffee were dutifully noted in the local press, including the *Times,* on April 2 and 12, 1911.

Mulholland's condemnation of "capitalists" was contained in an interview with the *Los Angeles Record* (March 22, 1912) and is often referred to (Kahrl, *Water and Power,* 190; Catherine Mulholland, *WM&RLA,* 211; Ostrom, *Vision or Villainy,* 161). The comment was given currency by McCarthy in "Water," February 12, 1938, page 30, and is in keeping with his disdain for speculators and special interests, though he was just as impatient with idealists who could not grasp the fact that wherever the aqueduct terminated, someone was bound to own the land nearby.

A thoughtful reassessment of the work of the AIB is found in Hoffman, "The Los Angeles Aqueduct Investigation Board of 1912," pages 329–360.

The quote from *The Coming Victory* (November 25, 1911) is cited by Ostrom, *Water & Politics,* pages 56–57.

A summation of the AIB findings is found in Ostrom, page 58.

17. IF YOU DIG IT, THEY WILL COME

"Thoughts on the Aqueduct Controversy" is the title of chapter 29 of the original Catherine Mulholland manuscript, "William Mulholland and the Making of Los Angeles."

A truncated discussion of Mulholland's family life is found in Catherine Mulholland, *WM&RLA,* page 374. Such materials are also found in greater detail throughout her unpublished, original manuscript.

Mulholland's comment that he was in need of a long rest is from the *Los Angeles Record,* March 22, 1912.

Mulholland's rosy assessment of the work to date comes from the *Los Angeles Times,* May 17, 1912.

Details of the Clearwater Tunnel explosion are from the *Times*, June 17, 1912.

Mulholland's response to critics was carried by most papers, including the *Times*, July 19, 1912.

The laborer's death was noted by the *Times* on June 25, 1912.

For Lippincott's summary of progress see the *Times*, August 10, 1912.

Lippincott had earlier provided a meticulous summary of cost-cutting measures throughout the project to reporters; see the *Times*, January 30, 1912.

"Exuberant" might be the term characteristic of accounts of most proposed celebrations; see the *Times*, July 30, 1912.

The "high line" and Pasadena's prospects were discussed in the *Times* on December 28, 1912.

The attorney's opinion and the election were announced in the *Times* on December 27, 1912.

Mulholland's letter to the Public Service Commission was reprinted in the *Times* on April 13, 1913.

18. LAST MILE

The *Inyo Independent* carried the story of the disaster on January 31, 1913; it was also reported by the *Los Angeles Times* on July 19, 1912.

Ratich's story was reported by the *Times* on August 15, 1912.

Taylor's summary of project casualties, and more, is found in *Men, Medicine & Water*, pages 162–163. The accident at the cement plant and subsequent is found on pages 143–144.

The decision to reschedule the bond election was noted in the *Times* on January 30, 1913.

Mulholland's letter to Willard is detailed in Catherine Mulholland, *WM&RLA*, pages 230–231.

The ceremony attendant to the formal opening of the gates at the diversion point was widely covered. See the *Times*, February 14, 1913.

Voting tallies are from the *Times*, April 16, 1913.

Rose's turnabout is described in Ostrom, *Water & Politics*, page 59.

The Sand Canyon blowout is described by Heinly, "Failure of the Sand Canyon Pressure-Tunnel Siphon," and by Mulholland in his "Complete Report," page 26.

19. CASCADE

The turning of the waters back into the aqueduct was covered by the *Los Angeles Times* on September 27, 1913.

Mulholland's heartening report on water flow in the aqueduct is from the *Times*, October 2, 1913.

Mulholland's "firecracker" quote is from Catherine Mulholland, *WM&RLA*, page 242.

The mention of the report on Lillie's condition is from Nadeau, *The Water Seekers*, page 62.

The event was, of course, widely covered in Los Angeles newspapers, including the *Times*, November 6, 1913. Mulholland's statement regarding his conviction that all would be forgiven once the waters arrived is from the *Times*, November 16, 1913.

Nadeau reports Mulholland's comment on Mathews keeping him out of jail, page 44.

Shaw's comments were carried by the *Times* in advance of the celebration, September 4, 1912.

A final summary of the aqueduct's features is found in Osborne, "Completion of the Los Angeles Aqueduct," page 271, and in the "Complete Report," page 263.

Dollar figures are from the *Times*, June 22, 1912, and the "Complete Report," page 262.

The interview with Mulholland is by Alan Hughes, the *Times*, November 9, 1913.

20. IN THE SHADE OF ACCOMPLISHMENT

LADWP's Fred Barker is the source of the actual date that Owens River water finally reached customer taps.

The collapse of the line in the Antelope Valley was reported by the *Los Angeles Times* on March 12, 1914, and in the "Complete Report," page 21.

Mulholland's departure for Berkeley was covered by the *Times* on May 12, 1914.

Mulholland's testimony was recounted by the *Times* on January 30, 1915.

The three-dimensional map was reported on by the *Times* on July 4, 1915. The map and other details of his post-aqueduct activity may be found in Catherine Mulholland, *WM&RLA*, pages 256–259.

Discussion of post-aqueduct development is found in Tzeng, "Eastern Promises," pages 49–50, and Ostrom, *Water & Politics*, pages 161–167.

A summary of annexation activities is found in Ostrom, page 161.

Discussion of the Colorado River Project is in Ostrom, pages 168–174.

Mulholland's dealings with Eaton in regard to the Long Valley Reservoir are characterized in Nadeau, *The Water Seekers*, page 64, and Chalfant, *The Story of Inyo*, page 383.

The city's machinations regarding the dam site in the Long Valley are described by Ostrom, page 120.

Eaton's interview was published in the *Riverside Daily Press* on March 14, 1906.

Fred Eaton's view of Mulholland's motives was affirmed by Harold "Hal" Eaton in a 2014 interview.

The dig at the city's expense is reported by Nadeau, page 71.

Nadeau's is a stirring account of the contretemps at Big Pine, pages 73–74.

The *San Francisco Call* articles were widely reprinted and are summarized in Catherine Mulholland, *WM&RLA*, pages 286–287.

21. LET THE BOMBINGS BEGIN

News of the bombing was reported in Los Angeles papers, including the *Times*, on May 22, 1924.

The interview with Mulholland is from the *Times*, June 25, 1924.

The PSC report is detailed in the *Times*, August 2, 1924.

Hall's adventures and Mulholland's response were reported by the *Times* and later woven into Nadeau's colorful account of the so-called Owens Valley Water Wars of the 1920s. *Times*, August 30 and 31, 1924; Nadeau, *The Water Seekers*, pages 77–82.

Details of the deteriorating situation in the Owens Valley are found in Ostrom, *Water Politics*, page 123, as well as a report in the *Times* on September 4, 1924.

A report of the meeting in the Valley is from the *Times*, September 6, 1924.

The counterproposals were reported on by the *Times* on November 17, 1924.

News accounts of the seizure of the Alabama Gates are from the *Times,* November 19 and 20, 1924.

The estimate of the amount of water turned out of the aqueduct by the protestors comes from LADWP's Fred Barker, who bases the figures on seasonal averages and an examination of news photographs of the time.

McClure's part is described by Ostrom, page 122.

A report on the pending resolution of the situation in the Owens Valley is from the *Times,* December 4, 1924.

The city's land ownership figures are from Ostrom, page 123.

Coverage of the May bombing and Mulholland's response was of particular interest to the Los Angeles papers, including the *Times,* May 13 and 15, 1926; and the *Express,* the *Examiner,* and the *Evening News,* all May 13, 1926.

Del Valle's statements are taken from a story in the *Times,* April 24, 1927.

Mulholland's comment was to a reporter for the *Times,* May 28, 1927.

Nadeau's gripping account of the collapse of the Watterson banks is found on pages 110–114 of *The Water Seekers.*

22. FAILURE

Mulholland's remarks on the Watterson banks were reported by the *Los Angeles Times* on August 7, 1927.

Detailed accounts of the completion and opening of the Mulholland Highway are found in the *Times,* October 19, 1924, and December 28, 1924. Lloyd Wright, son of Frank Lloyd Wright, created two versions of the Hollywood Bowl's band shell, but neither survived. The current shell, described as combining the superior acoustics of the Wright designs with visual elements of the 1929 structure, was completed in 2004.

Muholland's presentation on the Colorado was reported by the *Times* on March 1, 1928.

Storage capacity of the St. Francis Dam, along with other details, comes from Rogers, "A Man, a Dam and a Disaster," pages 20–21.

Concerns with the dam's height-to-base ratio as well as other design issues are detailed in Rogers, pages 27–32.

Details of the inspection trip are drawn from Mulholland's testimony at the coroner's hearing.

Mulholland's quote is taken from Catherine Mulholland, "William Mulholland and the St. Francis Dam," page 126.

Description of the route to the disaster is from the *Times*, March 14, 1928. Identification of Mulholland's driver is from Catherine Mulholland, "William Mulholland and the St. Francis Dam," pages 136–137.

Accounts of the disaster were widespread. Details here are taken in large part from the *New York Times*, March 14, 1928.

The Curtis and Rivera accounts are from McCarthy, "Water," page 37.

The Orton letter is reproduced in Catherine Mulholland, "William Mulholland and the St. Francis Dam," pages 128–129.

Statistics on family casualties, reparations paid, and the like are from the report of the Citizens Restoration Committee, July 15, 1929.

The pilot's report is from the *Times*, March 14, 1928.

Relief efforts are described in Matson, *William Mulholland*, pages 56–57.

Reparations figures are from the *Times*, March 20, 1928, and from the report of the Citizens Committee.

The arrangement with the contractors' association is reported by Matson, page 58.

Details of the investigations were reported by the *Times*, March 16, 1928.

Mulholland's request for leave and its denial appear in the *Times*, March 20, 1928.

The quote from Lillian Darrow is in Catherine Mulholland, "William Mulholland and the St. Francis Dam," page 130.

The "Kill Mulholland" sign is reported by Outland, *Man-Made Disaster*, page 58.

Mulholland's initial appearance at the coroner's inquiry was reported in the *Times* on March 22, 1928.

Mulholland's "fasten the blame on me" statement was reported by the *Times* on March 28, 1928.

23. FORGET IT, JAKE. IT'S CHINATOWN

William Bodine Mulholland's statement is from Catherine Mulholland's *WM&RLA*, page 327.

George Bejar's comments are from an interview conducted by Don Ray and William Thomas on April 20, 1980, and are included in Ray's forthcoming collection of first-hand accounts of the dam collapse aftermath, *Without Warning: Diary of a Disaster.*

On exculpation of Mulholland, see Rogers, "A Man, a Dam, and a Disaster," page 22 and passim.

The rejoinder to Rogers is from Jackson and Hundley, "Privilege and Responsibility: Wm. Mulholland and the St. Francis Dam Disaster." See especially pages 44–47 and passim.

The honor from the AAE was noted in the *Los Angeles Times,* April 10, 1921.

Testing on Mulholland Dam was reported by the *Times* on June 5, 1928.

The motel anecdote is found on pages 328–329 of Catherine Mulholland's *WM&RLA*. Details of Mulholland's passing are from Catherine Mulholland, *WM&RLA*, page 330, and Matson, *William Mulholland,* page 65.

The LADWP's Fred Barker points out that there was also a Liberty Ship named after Mulholland, as well as a Mulholland Tank and Mulholland Pumping Station.

The author's interview with Robert Towne took place on July 15, 2013.

The *Times* interview with Berry was published on October 30, 2013, presumably as part of the run-up to a celebration of the 100th Anniversary of the Aqueduct's Arrival in November 2013.

The anecdote regarding the LADWP official's response to *Chinatown* is related in Hoffman, *Vision or Villainy,* page xiii.

Mulholland's statement regarding the degree of his unfamiliarity with Otis is from his 1911 speech to the City Men's Club cited earlier.

Speculation concerning the role of Moses Sherman in the exercise of the land syndicate options in the San Fernando Valley in 1905 is that of the author.

24. CITY OF ANGELS

The quotations on the power of historical allusion in *Chinatown* are from Scott, "Either You Bring the Water to L.A. or You Bring L.A. to the Water."

Mulholland's distaste for vacationing is oft cited. This instance is from Nadeau, *The Water Seekers,* page 123.

The final meeting between Fred Eaton and Mulholland is described by Nadeau, page 131.

The author's interview with Harold "Hal" Eaton took place on April 5, 2014.

Mulholland's statements to his daughter Rose regarding Eaton, along with other details regarding Eaton's aspirations, are from Nadeau, pages 129–131.

The "little piety" quote is from McCarthy, "Water," March 26, 1938, page 31, as is the "big Los Angeles" quote found later in this chapter.

Mulholland's passing and the terms of his will were reported by the *Times* on July 26 and 30, 1935.

Figures concerning the city's acquisitions in the Owens Valley are from Ostrom, *Water & Politics,* page 127.

The note on chili pepper farming in the Owens Valley was reported in the *Times* on May 16, 2009.

The statistics regarding contemporary farming in the Owens Valley are from Vorster, "The Development and Decline of Agriculture in the Owens Valley," page 282.

The Lone Pine Highway 395 matter was reported on by the *Times* on May 19, 2009.

Catherine Mulholland's recollections of her family's furtive incursions into the Owens Valley are noted in *WM&RLA,* page 329.

The *L.A. Weekly* series by John Shannon was republished as "Fresh Meat for Bill Mulholland," in *Heritage.*

The quotation from Libecap is to be found on page 7 of a concise version of his argument, "Chinatown."

The mule team parade was featured, including a photo the likes of which Mulholland surely would have approved, by the *Times* on November 11, 2013.

The characterization of Mulholland Drive is in David Thomson, *Beneath Mulholland,* quoted by David L. Ulin, "There it Is. Take It," page 3.

DWP figures are from an interview with DWP spokesman and engineer Fred Barker on March 19, 2014.

President Obama's statements were reported by the *Times* on February 14, 2014.

Fred Eaton's remarks are reported in the *Riverside Daily Press,* July 7, 1892.

Mulholland's "Damn the non-reader" quote is from Prosser, "The Maker of Los Angeles," page 44.

The letter to the brick maker is described in a letter from H. W. Keller to the editor of *Pacific Saturday Night*, November 27, 1937, page 5.

Spriggs details the interview and her encounter with Mulholland and the oak on pages 69–70 of her thesis, "The History of the Domestic Water Supply of Los Angeles."

SELECTED BIBLIOGRAPHY

GOVERNMENT DOCUMENTS

City of Los Angeles Aqueduct Investigation Board. "Report of the Aqueduct Investigation Board to the City Council of Los Angeles." August 31, 1912.

Department of Public Works. "First Annual Report of the Chief Engineer of the Los Angeles Aqueduct to the Board of Public Works." March 15, 1907.

Los Angeles Board of Public Service Commissioners. "Complete Report on Construction of the Los Angeles Aqueduct." Los Angeles: Department of Public Service, 1916.

———. *Annual Reports,* 1917–1925.

Los Angeles Board of Water and Power Commissioners. *Annual Reports,* 1926, 1929.

Los Angeles, Citizens' Restoration Committee. "Report on Death and Disability Claims, St. Francis Dam Disaster in Los Angeles and Ventura Counties." July 15, 1929.

Los Angeles County Coroner. "Transcript of Testimony and Verdict of the Coroner's Jury in the Inquest over Victims of St. Francis Dam Disaster." Book 26902. Los Angeles Country Archives. March 21, 1928.

Los Angeles Department of Public Works. Los Angeles Aqueduct. *Annual Reports,* 1907-1912.

BOOKS

Chalfant, W. A. *The Story of Inyo.* Rev. ed. Bishop, CA: 1933.

Darrow, Clarence. *The Story of My Life.* New York: Scribner's, 1932.

Hoffman, Abraham. *Vision or Villainy: Origins of the Owens Valley–Los*

Angeles Water Controversy. College Station: Texas A&M University Press, 1981.

Kahrl, William L. *Water and Power: The Conflict Over Los Angeles' Water Supply in the Owens Valley*. Berkeley and Los Angeles: University of California Press, 1983.

LeConte, Joseph. *Elements of Geology: A Text-Book for Colleges and for the General Reader*. New York: D. Appleton, 1883.

Libecap, Gary. *Owens Valley Revisited: A Reassessment of the West's First Great Water Transfer*. Palo Alto, CA: Stanford University Press, 2007.

Matson, Robert William. *William Mulholland: A Forgotten Forefather*. Stockton, CA: Pacific Center for Western Studies, 1976.

Mayo, Morrow. *Los Angeles*. New York: Knopf, 1933.

McGroarty, John Steven. *History of Los Angeles County*. 3 vols. Chicago & New York: American Historical Society, 1923.

Mulholland, Catherine. *William Mulholland and the Rise of Los Angeles*. Berkeley and Los Angeles: University of California Press, 2000.

Nadeau, Remi A. *The Water Seekers*. Garden City, NY: Doubleday, 1950.

Nordhoff, Charles. *California: For Health, Pleasure, and Residence. A Book for Travelers and Settlers*. New York: Harper and Brothers, 1873.

Nunis, Doyce B., Jr., ed. *The St. Francis Dam Disaster Revisited*. Los Angeles: Historical Society of Los Angeles, 1995.

The 100 Most Important Americans of the 20th Century: Life Special Issue. New York: Time Inc. Magazine Company, 1990.

Ostrom, Vincent. *Water & Politics: A Study of Water Policies and Administration in the Development of Los Angeles*. Los Angeles: The Haynes Foundation, 1953.

Outland, Charles F. *Man-Made Disaster: The Story of St. Francis Dam, Its Place in Southern California's Water System, Its Failure and the Tragedy of March 12 and 13, 1928, in the Santa Clara River Valley*. Rev. ed. Glendale, CA: Arthur H. Clark, 1977.

Reisner, Marc. *Cadillac Desert: The American West and Its Disappearing Water*. New York: Penguin, 1993.

Spilman, W. T. *The Conspiracy: An Exposure of the Owens River Water and San Fernando Land Frauds*. Los Angeles: The Alembic Club, 1912.

Standiford, Les. *Last Train to Paradise*. New York: Crown, 2002.

———. *Meet You in Hell.* New York: Crown, 2005.
Steffens, Lincoln. *The Autobiography of Lincoln Steffens.* New York: Harcourt Brace and World, 1958.
Taylor, Raymond G. *Men, Medicine & Water: The Building of the Los Angeles Aqueduct, 1908-1913.* Edited by Doyce B. Nunis. Los Angeles: Friends of the LACMA Library and the Los Angeles Department of Water and Power, 1982.
Thomson, David. *Beneath Mulholland.* New York: Knopf, 1997.

ARTICLES

"California's Little Civil War." *The Literary Digest,* December 6, 1924.
Cross, Frederick C. "My Days on the Jawbone." *Westways,* May 1968, 3–8.
Hampton, Edgar Lloyd. "An Irishman Moves West." *Success,* August 1923, 28–31.
Heinly, Burt A. "Carrying Water through a Desert: The Story of the Los Angeles Aqueduct." *National Geographic,* July 1910, 568–596.
———. "The Failure of the Sand Canyon Pressure-Tunnel Siphon of the Los Angeles Aqueduct," *Engineering News-Record* 69, no. 23 (July 15, 1913): 1198–1200.
"History of Drinking Water Treatment." In *Twenty-Five Years of the Safe Drinking Water Act: History and Trends.* United States Environmental Protection Agency, February 2000.
Hoffman, Abraham. "Charles F. Outland, Local Historian." In *The St. Francis Dam Disaster Revisited,* edited by Doyce B. Nunis Jr. Los Angeles: Historical Society of Southern California, 1995.
———. "Joseph Barlow Lippincott and the Owens Valley Controversy: Time for Revision," *Southern California Quarterly* 54, no. 3 (Fall 1972): 239–254.
———. "The Los Angeles Aqueduct Investigation Board of 1912: A Reappraisal." *Southern California Quarterly* 62, no. 4 (Winter 1980): 329–360.
Hurlbut, W. W. "The Man and the Engineer." *Western Construction News,* April 25, 1926, 44.
Jackson, Donald C., and Norris Hundley, Jr. "Privilege and Responsibility: Wm. Mulholland and the St. Francis Dam Disaster." *California History* 82, no. 3 (2004): 8–47.
Keller, H. W. "Letters." *Pacific Saturday Night,* December 27, 1937, 5.
Libecap, Gary D. "Chinatown: Owens Valley and Western Water Real-

location; Getting the Record Straight and What It Means for Water Markets." *Texas Law Review* 83, no. 7 (June 2005): 2055–2089.

Lippincott, J. B. "William Mulholland—Engineer, Pioneer, Raconteur: Part I, His Start in Life and His Service in the Los Angeles City Water Company." *Civil Engineering* 2, no. 2 (February 1941): 105–107.

———. "William Mulholland—Engineer, Pioneer, Raconteur: Part II, The Owens Valley Acqueduct and Later Work." *Civil Engineering* 2, no. 3 (March 1941): 161–164.

"Management of Water Flowing through Pipes—Water Meters." *Scientific American* 23 (1870): 279.

McCarthy, John Russell. "Water: The Story of Bill Mulholland." *Pacific Saturday Night,* Chs. 1–15, October 30, 1937–March 26, 1938.

Moody, Charles Amadon. "Los Angeles and the Owens River." *Out West* 23 (October, 1905): 421–442.

Mulholland, Catherine. "William Mulholland and the St. Francis Dam." In *The St. Francis Dam Disaster Revisited,* edited by Doyce B. Nunis Jr. Los Angeles: Historical Society of Southern California, 1995.

Newell, W. H. "The Reclamation Service and the Owens Valley." *Out West* 23 (October, 1905): 454–461.

"New York City Watershed Retrospective," CatskillArchive.com.

"Nine Miles of Siphons" *The Literary Digest,* March 1, 1913, 452.

Osborne, Henry Z. "The Completion of the Los Angeles Aqueduct." *Scientific American* 109, no. 19 (November 8, 1913): 364–367, 371–372.

Prosser, Richard. "The Maker of Los Angeles." *Western Construction News,* April 25, 1926, 43–44.

Rogers, J. David. "A Man, a Dam, and a Disaster: Mulholland and the St. Francis Dam." In *The St. Francis Dam Disaster Revisited,* edited Doyce B. Nunis Jr. Los Angeles: Historical Society of Southern California, 1995.

Scott, Ian S. "Either You Bring the Water to L.A. or You Bring L.A. to the Water: Politics, Perceptions and the Pursuit of History in Roman Polanski's *Chinatown.*" *European Journal of American Studies* 2, no. 2 (Autumn 2007): Document 1.

Shannon, John. "Fresh Meat for Bill Mulholland: Or How I Learned to Love the Owens Valley Water Wars." *Heritage,* Winter 1991–92, 5–7, Spring 1992, 5–10.

Shrader, E. Roscoe. "A Ditch in the Desert," *Scribner's*, May 1912, 538–550.

Sklar, Anna. "The Man Who Built Los Angeles: Engineer, Mayor, Visionary and Forgotten Man." *Los Angeles City Historical Society Newsletter*, November 2013, 4–7.

Smythe, William B. "The Social Significance of the Owens River Project." *Out West* 23 (October 1905): 443–453.

Tzeng, Timothy. "Eastern Promises: The Role of Eastern Capital in the Development of Los Angeles, 1900–1920." *California History* 88, no. 2 (2011): 32–63.

Ulin, David L. "There It Is. Take It." *Boom* 3, no. 3 (Fall 2013).

Van Bueren, Thad M. "Struggling with Class Relations at a Los Angeles Aqueduct Construction Camp." *Historical Archaeology* 36, no. 3, Communities Defined by Work: Life in Western Work Camps (2002), 28–43.

Vorster, Peter. "The Development and Decline of Agriculture in the Owens Valley." In *The History of Water*, eds. Clarence A. Hall, Victoria Doyle-Jones, and Barbara Widawski. White Mountain Research Station Symposium 4. Los Angeles: White Mountain Research Station, (1992): 268–284.

W. S. B. "Record of the Owens River Project." *Out West* 30, no. 10 (April 1909): 258–277.

Widney, Erwin W. "We Build a Railroad." *Touring Topics* 23, no. 3 (March 1931): 36–41, 52–53.

THESES, UNPUBLISHED PAPERS, AND MANUSCRIPT COLLECTIONS

Catherine Mulholland Collection, Delmar Oviatt Library, California State University at Northridge.

Layne, J. Gregg. "Water and Power for a Great City." Typescript, Los Angeles Department of Water and Power, 1957.

Los Angeles Aqueduct/Department of Water and Power Papers, Eastern California Museum, Independence, California.

Mulholland, Catherine. "William Mulholland and the Making of Los Angeles". Archives and Special Collections, Oviatt Library, California State University, Northridge.

Mulholland, William. "Autobiographical Sketch." February 8, 1930. Courtesy Los Angeles Department of Water and Power.

———. "Suggestions for the Discussion of the Subject of the Disposal of the Water Supply from the Los Angeles Aqueduct." Office Files 1902–1914, Los Angeles Department of Water and Power.

Ray, Don. *Without Warning: Diary of a Disaster—Firsthand Accounts of the St. Francis Dam Disaster*, forthcoming.

Spriggs, Elizabeth M. "The History of the Domestic Water Supply of Los Angeles." Master's thesis, University of Southern California, 1931.

NEWSPAPERS

Goldfield News
Inyo Independent
Inyo Register
Los Angeles Evening Express
Los Angeles Examiner
Los Angeles Herald
Los Angeles Record
Los Angeles Times
New York Times
Riverside Daily Press
San Francisco Call
San Francisco Chronicle

INTERVIEWS

Fred Barker, Los Angeles Department of Water and Power, March 18 and May 29, 2014.

Harold "Hal" Eaton, Los Angeles, April 5 and May 28, 2014.

Christine Mulholland, San Luis Obispo, California, May 25–26, 2014.

Don Ray, Burbank, California, September 24, 2014.

Robert Towne, Los Angeles, July 15, 2013.

INDEX

ABOUT THE AUTHOR

LES STANDIFORD is the national bestselling author of twenty books and novels, including the John Deal mystery series, and the works of narrative history *The Man Who Invented Christmas*, a New York Times Editor's Choice, and *Last Train to Paradise*, the "One Book" Choice of more than a dozen public library systems. He is the founding director of the creative writing program at Florida International University in Miami, where he lives with his wife, Kimberly, a psychotherapist and artist. Visit his website at www.les-standiford.com.